HASSELL
POETIC PRAGMATISM

Pesaro Publishing

Editor: Patrick Bingham-Hall
Design: Patrick Leong

Pesaro Publishing
PO Box 74
Balmain NSW 2041
Sydney Australia
Phone 61 2 9555 7422
Fax 61 2 9818 6999
Email pesaro@bigpond.net.au

Powell, Robert, 1942- .
Hassell Architects: poetic pragmatism.

Includes index.
ISBN 1 877015 09 1.

1. Hassell Pty. Ltd. 2. Architecture–Australia.
3. Architecture, Modern–20th century. I. Bingham-Hall,
Patrick. II. Title.

720.994

First Published in 2003 by
Pesaro Publishing, Sydney, Australia

Photography © Patrick Bingham-Hall except as
otherwise credited
Text: © Robert Powell and Pesaro Publishing

All rights reserved. No part of this publication may be
reproduced, stored in or introduced to a retrieval system
or transmitted in any form or by any means, electronic or
otherwise, without the prior permission of the publishers.

Colour Origination by Universal Graphics, Singapore
Printed and Bound by Star Standard Industries, Singapore

Front Cover Image: Olympic Park Rail Station
                   Sydney, NSW, Australia 1996–1998

HASSELL
POETIC PRAGMATISM

ROBERT POWELL

PHOTOGRAPHY BY PATRICK BINGHAM-HALL

Contents

8 **Poetic Pragmatism**
Robert Powell

30 **Northern Metropolitan Institute of TAFE**
Heidelberg Victoria Australia

34 **Bankers Trust Sevice Centre**
Adelaide South Australia Australia

35 **ADI Ammunition Manufacturing Complex**
Benalla Victoria Australia

36 **Victoria's Open Range Zoo**
Werribee Victoria Australia

38 **River Torrens Linear Park**
Adelaide South Australia Australia

39 **St Vincent's Public Hospital**
Melbourne Victoria Australia

40 **Chifley Square and Café**
Sydney New South Wales Australia

44 **Tai Po Waterfront Park**
Hong Kong Peoples Republic of China

50 **Minter Ellison Offices**
Melbourne Victoria Australia

52 **Empire Theatre**
Toowoomba Queensland Australia

56 **Gordon Yu-Hoi Chiu Building**
University of Sydney New South Wales Australia

60 **Olympic Park Rail Station**
Sydney Olympic Park New South Wales Australia

70 **Canberra Playhouse Theatre**
Canberra Australia Capital Territory Australia

74 **Ohel Leah Synagogue**
Hong Kong Peoples Republic of China

78 **Millennium Parklands**
Sydney New South Wales Australia

82 **Qantas Domestic Terminal**
Sydney New South Wales Australia

88 **Commonwealth Law Courts**
Melbourne Victoria Australia

96 **Helvetica Café**
Melbourne Victoria Australia

98 **Alanbrae Residential Estate**
Melbourne Victoria Australia

99 **Sunshine Courthouse and Police Station**
Sunshine Victoria Australia

100 **Fox Studios Carpark**
Sydney New South Wales Australia

104 **Bent Street Entertainment Precinct**
Sydney New South Wales Australia

106 **Virgin Atlantic Clubhouse**
Hong Kong Peoples Republic of China

108 **Newington Housing**
Sydney   New South Wales   Australia

110 **Starcom Worldwide Offices**
Brisbane   Queensland   Australia

112 **David Jones Department Store**
Adelaide   South Australia   Australia

116 **Haymarket Public Spaces**
Sydney   New South Wales   Australia

118 **Roma Mitchell Arts Centre**
Adelaide   South Australia   Australia

126 **Riverbank Promenade**
Adelaide   South Australia   Australia

130 **Jones Lang LaSalle**
Melbourne   Victoria   Australia

134 **Hall Chadwick Centre**
Brisbane   Queensland   Australia

140 **Swiss Re Offices**
Sydney   New South Wales   Australia

142 **North Sydney Olympic Pool**
Sydney   New South Wales   Australia

150 **Green Garden Clubhouse**
Guangzhou   Peoples Republic of China

152 **London Mission Clubhouse**
Hong Kong   Peoples Republic of China

154 **Jacobs Creek Visitor Centre**
Barossa Valley   South Australia   Australia

160 **Gadens Offices**
Brisbane   Queensland   Australia

164 **Santos Petroleum Engineering Building**
University of Adelaide   South Australia   Australia

168 **Tognini's Salon**
Brisbane   Queensland   Australia

170 **Margate Foreshore Redevelopment**
Brisbane   Queensland   Australia

171 **Maroubra Beach Foreshore Promenade**
Sydney   New South Wales   Australia

172 **Goldfield Green Town**
Beijing   Peoples Republic of China

174 **HASSELL Sydney Office**
Sydney   New South Wales   Australia

176 **Allens Arthur Robinson Offices**
Melbourne   Victoria   Australia

180 **Victoria and Tote Parks**
Sydney   New South Wales   Australia

182 **The National Institute of Dramatic Art**
Sydney   New South Wales   Australia

192 **Awards**

200 **Acknowledgements, Photography Credits**

POETIC PRAGMATISM

ROBERT POWELL

"If there is one certainty in the postmodern conception of architecture it is that it is no longer about only one kind of architecture – there are many kinds of architecture, even within one designer's or one office's output and even within one building."

Ian Borden, "Revolution", in Blueprint, July/August 1999, pp37–38.

HASSELL's architecture defies immediate labelling. Given the wide geographical spread of the practice, the diverse cultural contexts and, not least, the different authorship of projects, the practice's output is both a manifestation of the richness and complexity of contemporary Australian architecture and a reflection of a world beyond Modernism which celebrates difference and pluralism.

A number of architectural writers have used the analogy of urbanised Australia as an archipelago of conurbations, with the major cities depicted as islands in the vast landscape with radically different climates, topography and, inevitably, architectural cultures.[1]

Many HASSELL projects exhibited in this book convey an architectural language that is sophisticated, elegant, relaxed and informal. The Olympic Park Rail Station is a sensuous, lyrical structure with an elegant folded roof and an interior that is all organised tidiness in its grey suit. The Qantas Domestic Terminal in Sydney has a similar calm and soothing ambience quite unlike the frenetic activity often associated with this building typology elsewhere in the world. In part it is the result of the muted and seductive use of colour – grey steelwork and blue-green glass – but it is also to do with the fluent rhythm of the roof structure and the elegant detailing.

The Commonwealth Law Courts in Melbourne have an altogether different intellectual appeal, with a witty and intelligent complexity behind a syncopated facade. Chaos is resolved into a heroic form. And the North Sydney Olympic Pool exhibits another sort of tension with the juxtaposition of hi-tech and existing Art Deco detailing. The outcome is a brilliant building with a supercharged atmosphere – all muscle and energy alongside intellect and elegance.

A similar combination can be found in Sydney's stylish Chifley Square where cosmopolitan chic is the order of the day. A public function – a cafe – within a predominantly corporate space is resolved in a coherent manner. It is both eloquent and reserved yet approachable. These qualities are also to be found in the Jacobs Creek Visitor Centre – an object in the landscape that is modest and unassuming yet

Opposite: National Institute of Dramatic Art (NIDA) Sydney, NSW, Australia 1999 – 2002.

[1] Philip Goad in *New Directions in Australian Architecture*, Pesaro Publishing, Sydney, 2001 refers to other writers who have developed this analogy including Phillip Drew in *The coast dwellers: Australians living on the edge*, Ringwood, Victoria: Penguin, 1994; Melbourne architect Alex Selenitsch; and Jennifer Taylor in *Australian Architects since 1960*, Canberra, RAIA Education Division, 1986.

Opposite: Roma Mitchell Arts Centre Adelaide, SA, Australia 1999–2000.
Below: Jacobs Creek Visitor Centre Barossa Valley, SA, Australia 2001.

distinctive and stylish. The same qualities are also apparent in the Gordon Yu-Hoi Chiu Building on the Sydney University campus.

Other HASSELL projects are exuberant and dramatic, even flamboyant. The soaring foyer of the National Institute of Dramatic Art (NIDA) in Sydney, dramatically lit by natural light from above during the day, is theatrically illuminated at night, and is complemented by the rich materiality and intimacy of the auditorium. The Dixon Street Mall in Sydney's Chinatown is effervescent, frivolous and sociable with a combination of charm and edginess. The classic Art Deco style Empire Theatre in Toowoomba is a slightly ostentatious building that HASSELL has extended without losing or compromising the playfulness and delight of the original.

In sharp contrast there is the tough brutalist vocabulary of the Northern Metropolitan TAFE at Heidelberg. Here the aesthetic is gritty, rugged and almost austere. A similar language is employed on the Sunshine Courthouse and Police Station. It has a serious, indeed stern, appearance, an aesthetic that is slightly softened by the curved, and ironically, welcoming form of the entrance. The Roma Mitchell Arts Centre in Adelaide also employs everyday materials—red brick, profiled steel and off-form concrete—to produce an object that is friendly and eloquent yet simultaneously tough, hard edged and resilient.

HASSELL's architecture in Asia embraces equally diverse and poetic languages expressing multiple meanings. Tai Po Waterfront Park in Hong Kong is a contemporary interpretation of a park born out of a society in transition. The park has a unique character reinforced by the exploratory use of softwood timber, concrete, steel and glass. The design questions preconceptions of what an urban park is and what it should be, in an international city undergoing momentous change. Elsewhere in Hong Kong the sensitive restoration of the Ohel Leah Synagogue gently affirms the presence of the Jewish diaspora, who have had a community in Hong Kong for more than 100 years.

But HASSELL's architecture is simultaneously deeply rooted in a pragmatic approach to design concerned with solving the functional relationships in the programme. The architecture is also driven by other imperatives, chiefly the growing global concern for sustainability and the protection of the planet's fragile ecosystems. This emerging sensibility towards ecology is the inspiration for projects such as the Hall Chadwick tower in Brisbane which incorporates a wide range of energy conserving technologies and has been acclaimed as the first real energy efficient office building in Queensland.

The design of the North Sydney Olympic Pool was similarly driven by a commitment to sustainability. It is the largest new public building in Australia incorporating 'green' technology. Close examination of projects such as the Jacobs Creek Visitor Centre reveals an ecological agenda—an interface of the poetic and the pragmatic can be observed in all of HASSELL's work illustrated in this book.

The Grid and the Landscape

HASSELL's architecture often arises at the intersection of, and out of an engagement with, the formality of the orthogonal planning grid in the major Australian cities. The practice takes the spatial and formal interests of the early Modernists, and shifts them from the notion of isolated objects in the landscape to become responsive artefacts in urban settings.[2]

As Philip Goad has observed, the late 20th Century Australian city has no counterpart in Great Britain or Europe. The images that spring to mind are those of cities like Vancouver, Toronto and Philadelphia. In these New World cities the grid was invariably used as an organising device by the city planners.[3]

Many of HASSELL's projects are located at the intersection or 'edge' of a city's formal grid and open space, thus they are simultaneously informed by the constructed and the natural landscape. This is true of the Commonwealth Law Courts in Melbourne, which engages with the grid and the open urban space of Flagstaff Gardens. Similarly the Roma Mitchell Arts Centre in Adelaide is embedded within the city grid, yet opens onto Light Square. Its distinctive form is an intelligent response to this location, and extends an invitation to the public.

The Riverbank Promenade in Adelaide is at the interface of the city grid and the Torrens River Valley, and successfully reconnects the city with the river from which it had gradually become estranged. The Canberra Playhouse also mediates with a formal plan—the City Beautiful layout of Walter Burley Griffin. The form of the building relates to the axial composition of the Civic Square and Vernon Circle.

In an even more dramatic manner, the North Sydney Olympic Pool engages with the urban grain at Milsons Point and the magnificent view of Sydney Harbour. The building clings resolutely to its roots and enjoys panoramic views of the waterfront. All of these buildings have a duality in their plan form.

The strength of HASSELL's architecture lies in the manner that the practice has recognised and grasped these opportunities at the edge of the grid in the urban context. This knowledge was turned to advantage when HASSELL was commissioned for the Olympic Park Rail Station, which connects with the gridded layout at Homebush, and creates the major public open space at the Olympic site.

Other projects are deeply embedded within the urban grid, and here HASSELL exploit the opportunities created by shifting geometry. This is the case with Chifley Square in Sydney, where the strong spatial composition is a direct response to the curious geometry of a former tram terminus at the junction of Hunter Street and Elizabeth Street.

But while much of HASSELL's work is located in the urban context, the practice has simultaneously nurtured a deep understanding of the Australian landscape. HASSELL was among the first practices in Australia to adopt an approach to architecture and planning that is sensitive to ecology. The special relationship of the Jacobs Creek Visitor Centre to the gentle slopes of the Barossa Valley was the outcome of HASSELL's exhaustive search for

Above: Canberra Playhouse Theatre
Canberra, ACT, Australia 1998–1999.

Opposite: Commonwealth Law Courts
Melbourne, Victoria, Australia 1995–1998.

---

2 Ken Maher in correspondence with the writer, 29th January 2003.
3 Philip Goad in *New Directions in Australian Architecture*, Pesaro Publishing, Sydney, 2001.

Above Top: National Insurance Company of New Zealand Adelaide, SA, Australia 1962.
Above: Balm Paints, Rocklea Queensland, Australia 1966.
Opposite Top: ANZ Bank Adelaide, SA, Australia 1956.
Opposite Middle: David Jones Department Store Adelaide, SA, Australia 1960.
Opposite Bottom: The Weir Restaurant Adelaide, SA, Australia 1968.

'the right site'. The practice was able to direct the client to a location with spectacular views, thus enhancing the attraction of the Centre.

In a different context, Victoria's Open Range Zoo replicates the savannah habitats of Africa, Australia and Asia, and has been integrated with skill and sensitivity into grasslands, 31 kilometres south-west of Melbourne. The Zoo educates the public in the importance of ecosystems and conservation of endangered species.

Another project by HASSELL, the Millennium Parklands Concept Plan, undertaken in collaboration with Peter Walker and Bruce MacKenzie, is one of the most inspiring legacies of the Sydney 2000 Olympic Games. It has provided recreational and educational experiences derived from a diversity of distinctly Australian natural and cultural settings. By restoring the important ecological resources of Homebush Bay's 'lowlands' into functioning waterways, a strong conservation message has been conveyed to Australia and to the world.

The First Generation

Adelaide is a conservative city, and it is challenging in the face of this conservative culture to erect avant-garde buildings. Nevertheless the city has given birth to three of Australia's largest architectural practices, all of whom have subsequently expanded from their roots in the South Australian capital.[4] This has been attributed to the very rigours of practice in Adelaide where to produce an architectural work of distinction requires unusual tenacity.[5]

The founding partners of the firm now known as HASSELL were Philip Claridge, Jack McConnell (a graduate of University of Melbourne in 1935) and Colin Hassell (an Adelaide University graduate of 1934). Hassell and McConnell, now retired, are living in Adelaide. In 1937 they were invited by Philip R Claridge to take part in a design competition for the Bank of New South Wales office in Adelaide (1938), which they won.[6] The Modern/Art Deco design reflected the most up to date thinking in Australia at the time – and the beginnings of the International Modern Movement in Adelaide. Subsequently they entered into partnership as Claridge, Hassell & McConnell in 1938. When Philip Claridge retired in 1949, the name of the practice became Hassell and McConnell.

From the outset, both Hassell and McConnell were committed to Modernism. McConnell trained in Melbourne at a time that saw the emergence of Australian Modernism in the work of Norman Seabrook (MacRobertson Girls High School 1934), and Stephenson and Meldrum (Mercy Hospital 1934–36 and Freemasons Hospital 1936). Seabrook, who was a studio master in the University of Melbourne atelier directed by Leighton Irwin, encouraged McConnell's interest in European Modernism.[7]

Colin Hassell won an Orient Line travelling scholarship, which enabled him to work in London from 1935 to 1937 – a time when many of the major figures of the Modern Movement such as Erich Mendelsohn, Serge Chermeyeff and Berthold Lubetkin were in practice in the UK.[8] He also travelled widely in Europe, and

---

4 HASSELL is placed 5th in the Top 10 Australian design consultancies. World Architecture magazine No. 112, January 2003.
5 Mariano de Duonni, a Principal in the Adelaide office of HASSELL in discussion with the writer, 1 November 2002.
6 The Bank of New South Wales office in Adelaide (1938) is now known as the Westpac Building.
7 Michael Page, Sculptors in Space: South Australian Architects 1836–1986.
8 Michael Page, ibid.

was impressed by the work of Willem Dudok, particularly Hilversum Town Hall (1927–1931).[9]

Jack McConnell's Deepacres housing project in Adelaide (1941) indicates a trenchant Modernist position. It is a simple white-painted, flat-roofed, box-like building of two and three storey residences arranged around three sides of a landscaped court–sharply reminiscent of work by early Modernists in Europe.

'The 1st generation' of the firm was driven by an architectural agenda dedicated to the reworking of International Modernism in the Australian context, and the firm built up a solid base of mainly industrial clients. In 1947 the practice opened an office in Melbourne under the direction of Henry Hayward. The first major job in the new office was a production facility for the International Harvester Co. at Dandenong in 1948. An example of the excellent work of the practice in this period was the HJ Heinz Offices and Factory (1952, completed in 1956). Further partners were introduced in 1957, and the practice was renamed Hassell McConnell and Partners.

It was a time of strong economic growth in Australia, and industrial buildings produced by the practice at this time included WD & HO Wills Factory in Adelaide (1955), Kodak Offices in Adelaide (1955), ANZ Bank in Adelaide (1956, utilising a gridded facade reminiscent of Arne Jacobson's Jesperson Offices in Copenhagen), Gerard Industries Ltd (1957), CSR Bacchus Marsh Offices and Workshops (1960), David Jones Department Store (1960), offices for the National Insurance Company of New Zealand (1962), International Harvesters Factory and Offices (1965), and BALM Paints HQ in Queensland, which won the RAIA Queensland Building of the Year award in 1966. All displayed an elegant Modern architectural language. The Weir Restaurant in Adelaide (1968) was another exquisite Modern building.

There was an influx of educational clients in the early 1960s, and the practice produced a number of seminal buildings, including the Flinders University North and South Ridge Precincts (1966 and 1968), which have echoes of Le Corbusier's language. In 1968 Ansett developed the new Channel 10 Television HQ in Melbourne. The award winning design by Henry Hayward led to other commissions from Ansett, and subsequently led to several airport projects for Ansett and Qantas which stimulated HASSELL's regional growth.

A Collaborative Practice

John Morphett graduated from the University of Adelaide in 1955 and joined the practice the same year. Shortly afterwards he departed for graduate studies at Massachusetts Institute of Technology (MIT). After graduation he worked with The Architects Collaborative (TAC), the practice founded in the USA by Walter Gropius, before moving in 1960 to the Rome office of TAC where he worked on the design of Baghdad University.[10]

Morphett rejoined the Adelaide practice in 1962 and was made a partner in 1967. But he would have to wait until 1970, following the retirement of Jack McConnell, before he could instil Gropius's ideas on interdisciplinary collaboration into the practice. Morphett, who headed "The 2nd generation" of partners, was

9 Colin Hassell in discussion with the writer, 3rd November 2002. Dudok's work influenced other architects in Australia. See for example the Heidelberg Town Hall (1937) by Peck and Kempler in association with A C Leith and Associates.

10 Walter Gropius was the founder, in 1919, of the Bauhaus where he worked in close collaboration with architects and artists such as Moholy-Nagy, Marcel Breuer, Wassily Kandinsky and Paul Klee.

Above Top: Adelaide Festival Centre 1968–1975.
Above: Rundle Street Carpark 1975–1977.
Opposite: 'M' on the Bund Restaurant Shanghai, Peoples Republic of China 1999.

totally committed to this approach and to Modern architecture. Several of his important projects can be seen in the South Australian state capital.

The firm were the designers of the Adelaide Festival Centre (1970) on the bank of the Torrens River. The distinguished architectural critic Jennifer Taylor has commented upon similarities with the Sydney Opera House.[11] It is not as ambitious, nor profound, as its Sydney counterpart but was nevertheless a successful design consisting of a Festival Theatre, the Dunstan Playhouse Theatre, the Space Theatre and an open-air amphitheatre seating 1200. It remains a lively arts and cultural space.

Morphett designed the steel-framed Rundle Street Car Park at the corner of Pulteney Street and Rundle Mall. It evokes memories of the curved facade of Erich Mendelsohn's Petersdorf Store in Breslau (1927) which employed what Reyner Banham has described as "structural assemblies of geometrical simple units which presented themselves to the eye as tidily profiled edges". The Rundle Street building has a strong horizontal emphasis, but this reading was later compromised by the city mayor's insistence that vertical bars be inserted between the floors as a safety measure.

In 1970 McConnell was awarded the Gold Medal of the Royal Australian Institute of Architects for his contribution to the development of architecture in Australia. He had been President of South Australia Chapter of the RAIA in 1966, and in 1968 he was elected National President of the RAIA for a two-year term.

Meanwhile, the second generation of the firm embraced Morphett's ideas on collaboration, and other disciplines were brought into the practice. They were less conventionally architecture-focused, and in 1972 the scope of the practice expanded into landscape architecture, and the firm also added a planning arm to provide services in urban and regional planning and environmental impact assessment. The practice subsequently won such projects as the planning of Adelaide Botanic Gardens and the Dandenong Flood Plain Study in Victoria.

In 1972 an office was opened in Sydney. Later an office would be opened in Auckland (1987) and others in Hong Kong (1990), Brisbane (1993) and Bangkok (1994). But the practice's focus on growth was arguably at the expense of their cutting edge in architectural design.

The Third generation

"Good architecture is an architecture of resistance. Tenacity and determination to succeed is essential".

Ken Maher November 2002

In the 1980s and early 1990s the focus of the practice shifted to the Melbourne office, headed by Tim Shannon, one of the 'third generation' of HASSELL partners. Shannon joined the practice in 1975, having previously worked with John Andrews, Australia's foremost exponent of brutalism, and studied urban design in Toronto.[12] Tim Shannon became Managing Director of HASSELL in 1990, and under his and John Morphett's leadership the

11 Jennifer Taylor, *Australian Architecture since 1960*, The Royal Australian Institute of Architects, Sydney, 1990.
12 Philip Drew, 'John Andrews: Australian Architecture's American Hero', *Architecture Australia*, May/June 2000.

practice expanded significantly with major commercial and institutional commissions in Melbourne. Shannon was also instrumental in seeking to renew the leadership of the practice and to chart a new direction in the 21st Century.

Christopher Wren, the Director in charge of the Brisbane office, trained at the Department of Landscape Architecture and Planning at the University of Pennsylvania, a programme strongly influenced by Ian McHarg, and joined HASSELL in the Adelaide office in 1975 to develop the landscape architecture practice. Stephen Williams was also recruited at this time to develop the planning practice. McHarg's ideas on ecological analysis, contained in his influential book 'Design with Nature' are the basis of much of HASSELL's planning and landscape work.[13]

The third generation partners were responsible for the development of the Hong Kong office. Initially the focus was upon landscape architecture, but this has subsequently been expanded into architecture and interior design. The Asian presence commenced with the 1990 merger of HASSELL's landscape team and EBC Pty Ltd, a landscape architecture practice with a successful office in Hong Kong, at that time under the leadership of Robin Edmond and Geoffrey Rex. More recently, Peter Duncan has guided the expansion of HASSELL into China, together with Terry Fan. Particularly fine projects are 'M' on the Bund, an elegant international restaurant in a 1930's heritage building in Shanghai (1999) and Tai Po Waterfront Park in Hong Kong (1995).

In 1983, Ken Lloyd, a founding partner of Inarc, joined HASSELL to lead the interior design practice. Following a successful period of growth there are now interior design teams under the leadership of Kirsti Simpson in all Australian offices, as well as in Shanghai, Hong Kong and Bangkok. Ken Lloyd subsequently relocated to Bangkok, and has led the expansion of activities in Thailand.

In 1995 Ken Maher, who operated his own successful practice in Sydney, was invited to merge his practice with HASSELL to develop the design culture and profile of the Sydney office. Over the following years the Sydney office expanded and this was paralleled by an increase in the volume of work being carried out in Asia. An office was subsequently opened in Manila in 1997.

Maher's architectural influences were Modernists, one crucial mentor being his tutor at the University of New South Wales in the 1960s, W.E. (Bill) Lucas (1924–2001). Bill Lucas's own influences were Amancio Williams and Rafael Soriano, and in this respect he was like other Australian Modernists such as Neville Gruzman, who were outside the mainstream Modernism represented by Harry Seidler. Lucas, unlike many of his contemporaries, "never lost sight of the architect as a maker of prototypes and always saw the universal possibilities in even the most modest of commissions".[14]

It was an important era in Sydney when several significant Modernist buildings were produced which advocated the integration of architecture with the landscape, and adopted a brutalist position in regard to materials. Lucas was ahead of his time and his work was "at the edge of the possible".[15] A Sydney sensibility

---

[13] Ian L. McHarg, the respected Scottish-born former Professor and Head of the Department of Landscape Architecture and Planning at the University of Pennsylvania is the author of *Design with Nature*, Doubleday/ Natural History Press, New York, 1969.

[14] Peter Myers in an obituary for Bill Lucas in 2001. Myers identifies Lucas's own mentors as the radical Argentinean architect Amancio Williams (1913 – 1989) and Rafael Soriano (1970–) the Greek Modern architect active on the West Coast of the USA.

[15] Peter Myers, ibid.

developed that would strongly influence Maher's generation, including Peter Myers and Richard Leplastrier. The Lucas House at Castlecrag, which was dubbed 'The Glass House', became a cult object for Sydney architects.

In the early 1970s Ken Maher (together with two colleagues from his graduating class at the University of NSW – Craig Burton and Colin Stewart) was runner-up to Richard Rogers and Renzo Piano in the international competition for the design of the Centre Pompidou in Paris.[16]

Maher continues to build upon a Modernist position. He is most passionately interested in work in the public domain, informed by postgraduate studies in landscape architecture and environmental studies. This is most powerfully illustrated in the Olympic Park Rail Station, the North Sydney Olympic Pool and Chifley Square.

A strictly functional building, without excess or ostentation, the Olympic Park Rail Station was a collaboration with Rodney Uren, a Melbourne principal of HASSELL and former director of Foster and Partners in the UK. It is "a significant piece of architecture which possesses a sensuousness rarely present in a public building".[17] It has been carefully crafted and there is a sense that the design concept has been rigorously followed through in the detailing. A 'sense of arrival' is celebrated in the tradition of all great railway stations.

Within the North Sydney Olympic Pool, fragments of the past have been retained. The textured brickwork, the plasterwork and ceramic tiles incorporating images of green frogs and leaping fish – all part of the 1936 Art Deco swimming pool – have been kept intact. In sharp contrast, a Miesian inspired extension reflects high technology. Materials are refined and calm, in contrast to the idiosyncratic textures of the old building. Routes are defined with remarkable clarity.

Chifley Square incorporates a café, a linear sculptural composition of grey granite and glass with a folded plane zinc-clad roof. A semi-opaque glass wall defines the south side of the restaurant. The 1.2 metre thick 'wall' accommodates the kitchen, toilets and the furniture store. Thus anchored, the café opens out to a north-facing open space to capture the sun. The wall has been extruded and projected to the west as a prismatic sculptural object. Semi-opaque glass gives way to transparent glass fixed to a milled-finish stainless-steel frame. Within it are overlapping diagonal sheets of glass and there is a play of refracted light moving into infinity.[18]

This body of work in Sydney exhibits clarity and legibility, ordering through structure and a sensitive response to site. Through these projects, and in the fine work by Paul Katsieris, Rodney Uren and others, HASSELL has re-established itself in the front rank of Australian design consultancies. Design quality is once again high on the agenda, and the practice is attracting many top design graduates. By the turn of the Millennium HASSELL could claim, on the basis of a substantial body of excellent new work, to be among the front-runners in the profession.

In 2000, John Morphett, now retired, was awarded the Gold Medal of the Royal Australian Institute of Architects.

Opposite: Olympic Park Rail Station
Sydney, Olympic Park, NSW, Australia 1996–1998.
Below: Chifley Square and Café
Sydney, NSW, Australia 1995–1997.

16 Ken Maher, Craig Burton and Colin Stewart, all from the same graduating class at the University of New South Wales – were runners-up to Richard Rogers and Renzo Piano in the Centre Pompidou competition (1972).

17 Robert Harley, 'Sulman awarded to Sydney Olympic Station', 1998 NSW Architectural Awards, *The Australian Financial Review*, June 18th, 1998.

18 Peter Ward, 'Sydney's 12-point plan', *The Weekend Australian*, December 13th – 14th, 1997.

Opposite and above: Tai Po Waterfront Park
Hong Kong, Peoples Republic of China 1992–1997.

Theoretical Underpinnings

HASSELL is committed to the continuation of Modernism in the context of changing societal values. In the 1960s the firm's architecture was straight-down-the-line Modernism aligned with Cartesian geometry. In the late 1980s it lost clarity of direction, as the practice expanded and grew, but in the mid 1990s it re-engaged with its roots, reinvented itself and re-emerged as a leading practice operating in a world beyond the strictures of Modernism.[19]

Tim Shannon notes that there is a consistent set of ideas that have evolved in relation to changes in society. Discussing the ideological and philosophical underpinnings of the practice he, like John Morphett, invokes a collaborative ethos in seeking a future direction for the firm. Collaboration respects differences and embraces pluralism.[20]

Ken Maher concurs, "The concept of collaboration and the multidisciplinary nature of the practice is fundamental," he says. "It explains not only the evolution of HASSELL, but also the way in which the work from a number of authors can be held together."[21]

HASSELL's corporate pyramid is relatively flat. Ideas are accepted from whichever level they are generated. It is possible for a designer to bring his or her own vision within the framework of a large practice, guided by the input of the design principals. For the gifted designer, HASSELL offers many opportunities that cannot be found in the offices of celebrated 'boutique' practices where design is dominated by one principal. There is a constant dichotomy in architecture between the avant-garde and the corporate firms. The avant-garde firms have to embrace corporate practices in order to carry out larger commissions, and corporate firms must be at the cutting edge of design in order to succeed in the market-place.[22] HASSELL seeks to draw these strands together, and are not short of excellent designers who operate at this 'cutting edge'.

The firm maintains close links with academia, and a number of the senior staff of HASSELL teach part-time. Ken Maher has led design studios over many years and is an Adjunct Professor at the University of New South Wales. Christopher Wren and Robin Edmond are also Adjunct Professors at QUT and RMIT respectively, and Tim Shannon, Paul Katsieris and Rodney Uren have taught design regularly.

HASSELL is a geographically diverse practice spread widely across markedly different climatic regions – subtropical, temperate and equatorial – and across markedly different cultural practices in these regions. The resultant architecture reflects these different cultural attitudes.

Christopher Wren, who established and developed the Brisbane office from the mid-1990s, ponders upon the differences between the output of the Brisbane office and the Melbourne office. "There are commonalities in the process, but not in the outcome. The further north you go in Australia there is a greater preoccupation with climate." [23] Whereas, as Philip Goad observes, "bereft of determining landscape and demanding climate, Melbourne's architects have always revelled in the artifice of architecture and formal experiment... Melbourne's pluralistic culture has an... often wayward and obsessive preoccupation with ideas".[24]

---

[19] Paul Katsieris in discussion with the writer, 4th November 2002.
[20] Tim Shannon in an address to staff in the Sydney office of HASSELL on 29th October 2002.
[21] Ken Maher in discussion with the writer 13th Nov. 2002.
[22] Paul Katsieris, ibid.
[23] Christopher Wren in discussion with the writer, 8th November 2002.
[24] Philip Goad, *A Short History of Melbourne Architecture*, Pesaro Publishing, Sydney, 2002.

Above: Northern Metropolitan Institute of TAFE Heidelberg, Victoria, Australia 1992–1994.

Opposite Top: Victoria Park Sydney, NSW, Australia 1999–2003.

Opposite Bottom: Goldfield Green Town Development Beijing, Peoples Republic of China 2001–2003.

HASSELL is not a practice driven by rhetoric or an overpowering ideology, nevertheless projects such as the Commonwealth Law Courts in Melbourne, the Metropolitan TAFE at Heidelberg and the Roma Mitchell Arts Centre in Adelaide reveal exploration of theoretical positions and phenomenological aspects of architecture.

The Commonwealth Law Courts is an appropriate building for Melbourne, which Katsieris sees as "a city of the mind in which beauty in its own right has no validity".[25] It is an intellectually challenging building, brimming with ideas. The architecture is thoughtful and vigorous. It has been welcomed by one writer/critic as an, "optimistic, free, lively and forward looking" building. The same writer concluded that it is, "an exercise at the edge of order and disorder that adds complexity and visual interest to what would otherwise be an assemblage of dumb boxes".[26]

The Northern Metropolitan Institute of TAFE at Heidelberg is markedly different and more in tune with the reductive structural expression seen in the work of the Sydney office. It is also strongly reminiscent of the "most famous Brutalist building", namely Hunstanton School in Norfolk, England, by Peter and Alison Smithson (1949–54), which was an "uncompromising exhibition of materials".[27]

The Roma Mitchell Arts Centre in Adelaide exhibits influences from the early university work of James Stirling at Leicester, while embracing current theoretical ideas propounded by Henri Lefebvre on "the architecture of the everyday". HASSELL takes 'ordinary' experiences, spaces and materials and transforms them into the 'extraordinary'.[28]

The Natural Environment and Ecological Influences

"We believe intelligent and ecologically responsive design will be necessary to meet the urban challenges ... of the 21st Century.

Tim Shannon November 1998.

Designing with respect for the natural environment and ecology is a constant and recurring theme in HASSELL's work. "HASSELL encourages everyone in the practice to have ideas," says Tony McCormick, a principal and key leader of the landscape architecture team. "But landscape architecture is what makes this company different." Landscape designers are involved early on in every project. "The land must tell you what to do," asserts McCormick. "This approach ensures that our projects have a connection with time and place, to ecology and hydrology."[29] McCormick and fellow director Ross de la Motte have been pivotal in refining HASSELL's research and analysis of natural systems and in establishing an ecological ethic.

The ecological basis of design is well illustrated in the River Torrens Linear Park (1975–1997), Australia's largest urban waterway rehabilitation project covering an area in excess of 500 hectares. The Torrens River was long regarded as a site for disposing of spoil, and this project has reversed the trend. Over 200 million cubic metres of earth have been removed for flood mitigation, and the river corridor has been stabilised with 1.5 million trees and shrubs.

Little Manly Point Park (1991), located adjacent to scenic Spring Cove on Sydney Harbour, is on a smaller scale but is equally

---

[25] Paul Katsieris, ibid.
[26] Joe Rollo, *Contemporary Melbourne Architecture*, UNSW Press, Sydney, 1999.
[27] Peter Murray & Stephen Trembly, *Modern British Architecture since 1945*, Frederick Muller Limited, London, 1984.
[28] H Lefebvre, *Architecture of the Everyday – The Everyday and Everydayness*, Princeton Architectural Press, Yale, 1997.
[29] Tony McCormick in discussion with the writer, 29th October 2002

impressive. A harbourside site, derelict until 1970, has been reclaimed for use as a public park. Some of the foundations of the former gasworks have been retained as a memory of the former industrial use, the site has been heavily planted with trees and shrubs, and a setting has been created for a variety of land and water-based recreational activities.

HASSELL was also principal consultant for the master plan of the 1100 hectare Desert Wildlife Park and Botanic Garden (1993) at Alice Springs. The park, straddling the MacDonnell Ranges, gives visitors an appreciation and understanding of the outback, which comprises seventy percent of Australia.

The message is also conveyed in HASSELL's master plan for Victoria Park in Sydney, which utilises an orthogonal grid dissected by a long crescent. Forty percent of the site is brought into the public domain, in Sydney's most advanced and environmentally sustainable water management system. The historic Tote building, a memory of the former racetrack, is restored and four parks are introduced – Tote Park, Woolwash Park, Nuffield Park and Joynton Park – each with its own special character. 800 trees and 70,000 new shrubs have been planted alongside roads and pedestrian pathways.

In a different context, the transformation of a 1.5 kilometre stretch of seafront in the Margate Foreshore Development in Queensland is remarkable, not so much for any grand architectural statement but because it transforms the ordinary and the everyday into something special. For a relatively modest financial investment of $A6 million, Redcliffe City Council have acquired an asset in the form of restored dunes and a boardwalk. They acknowledge this as a key component in their economic development and investment attraction scheme. In other words, good design pays.

HASSELL's projects in The Peoples Republic of China, such as the Sheung Wah Yuen Masterplan and the Goldfield Green Town Masterplan also make extensive use of water bodies in the landscape.

In each of these projects, HASSELL's design skills have made a huge contribution to the lives of ordinary people, regenerated degraded areas, and have suggested ways in which we can care for the embattled planet.

A Commitment to Sustainability

Closely related to ecological influences is a belief in 'green' buildings and an emerging commitment to sustainability. HASSELL is a foundation member of the Australian Green Building Council.

In the Gordon Yu-Hoi Chiu Building, a research facility on the Sydney University campus, HASSELL designed the main interior volume to be naturally ventilated. Cool air is drawn in at a low level along the northern and western façades and circulates up and through a narrow gap formed between the upper level of the office and the blade wall, by means of wind-driven ventilators. In addition a 'chimney-effect' is produced when the sunlight, through the roof level glazing, warms the blade wall causing the heated air to rise.

The design of the North Sydney Olympic Pool is also concerned with sustainability. The pool water for the new and existing swimming pools

Above: Gordon Yu-Hoi Chiu Building
University of Sydney, NSW, Australia 1997–1998.

Opposite: North Sydney Olympic Pool
Sydney, NSW, Australia 1997–2001.

is heated by a system that extracts heat from the waters of Sydney Harbour and from solar hot water panels mounted on the roof of the new indoor pool. This allows the outdoor pool to be used all year round.

The 22-storey Hall Chadwick office building in Brisbane is also driven by an agenda that stresses sustainability. The building makes extensive use of solar panels to generate electricity, and photovoltaic panels are integrated with the roof of the tower. The tenants use some of the solar power, and some is sold back to the grid

Urbanism, Urban Design and Planning

"Urban planning and urban design are increasingly important in the making of cities. The public domain is a critical aspect of much of our work."

Ken Maher 7th November 2002.

HASSELL has a keen interest in contemporary urbanism. Influences are drawn from recent European urban projects, while the practice also takes some of the spatial and formal interests of the early Modernists and shifts them from the notion of isolated objects in the landscape to become responsive artefacts in urban settings.

In Brisbane, a $A400 million development, master planned by HASSELL on the site of the former Gona Army Barracks and adjoining land owned by Queensland University of Technology (QUT), indicates the future direction of sustainable development in Australia. In 2001 HASSELL completed the plan for the 17 hectare Kelvin Grove Urban Village site, due for completion in 2008. The Master Plan for the Village lays down principles for Environmentally Sensitive Design (ESD), which minimises the use of fossil fuels, reduces car based travel, builds on the natural attributes of the area and provides low energy solutions, thereby reducing greenhouse gas emissions.

In January 2003 HASSELL was appointed as executive architects and designers of several of the key buildings for the $A1 billion Waterfront City Development project in Melbourne's Docklands. Features may include a waterfront square with a marina and a ferry, a market and an observation wheel.

These commissions come on the back of a number of highly acclaimed urban planning and urban design projects.

At Chifley Square, the skilful choreography of space by HASSELL gives coherence to the square and helps to define private space (the café), semiprivate space (the covered terrace), semi-public space (the shaded arcades around the square) and public space. The manipulation of structure, space, light and landscape ensures that the users' perception of the square changes from daytime to night-time.

At Fox Studios in Sydney, the urban design concept has integrated a number of heritage buildings. The development has been so successful that it has been declared a landmark precinct, and is now used internationally by the Australian Heritage Commission to illustrate Australian best practice in urban conservation and adaptive reuse. The success of the public domain is largely due to the urban designers and landscape architects in HASSELL who

Above: Doncaster Hills (project)
Melbourne, Victoria, Australia.

Opposite Top: Commonwealth Law Courts (project)
Adelaide, South Australia, Australia.

Opposite Centre: Macquarie Park Station
Parramatta Rail Link (project) Sydney.

Opposite Bottom: Liverpool-Parramatta Transitway
Station (project) Sydney.

answered a complex brief, requiring the integration of heritage building with a film culture which emphasises the transient marketable image.

Station Square, fronting the Olympic Park Rail Station, is the major public space at Sydney's Olympic Park. It adjoins the new Sydney Showground and links directly with Olympic Boulevard. This pedestrian hub is the experience that greets visitors on arrival as they emerge from the subterranean platforms, having travelled on the specially constructed 5.3 km loop off the main line from Sydney city centre. Similarly in Sydney's Dixon Street Mall, a multi-disciplinary project has had a profound effect on the area, enhancing its inherent attraction and sustaining a vibrant urban space pulsating with life. It is good for the city and good for the image of the city.

When John Morphett introduced the sculptural work of Otto Herbert Hajek and Bert Flugelman into the forecourt of the Adelaide Festival Centre in 1970, it essentially built upon the Bauhaus tradition of collaboration between architects and artists, and recalled Walter Gropius and Marcel Breuer's collaboration with artists Moholy-Nagy, Wassily Kandinsky and Paul Klee.

It is a tradition that HASSELL has continued in many of their landscape and architectural projects. In the Chifley Square project for example, the sculptural work of Simeon Nelson is introduced with an eight-metre high steel plate statue of Ben Chifley, and by the artist's very different Lightwall Crucimatralux structure, which is a design collaboration with HASSELL.

Similarly, in the Dixon Street Mall, HASSELL has collaborated with public artist Peter McGregor and lighting designer Barry Webb to create a vibrant overhead lighting display.

In the Victoria Park project, an intriguing water cascade designed by artist Jennifer Turpin and Michaelie Crawford is introduced into the public space at the heart of the development, while the façade of the David Jones Store in Adelaide carries the work of sculptor Catherine Trueman of the Gray Street Workshop

Design Focus

HASSELL is not being immodest when it claims to be far ahead of any other large commercial practice in Australia in terms of the integration of the different disciplines.[30] Moreover each discipline is a leader in its own respective field.[31]

HASSELL have, in the last decade, received an impressive number of awards for architecture, landscape architecture, interior design and planning. In 1998 the Olympic Park Rail Station was awarded the Royal Australian Institute of Architects Sir Zelman Cowen Award for the nation's best public building, and in the same year the building also won the Sir John Sulman Award for Outstanding Architecture presented by the New South Wales Chapter of the RAIA. This award had only been given three times in the previous decade.

HASSELL again won the Sir John Sulman Award from the RAIA (NSW Chapter) in 2002, this time for the National Institute of Dramatic Art (NIDA). The Jacobs Creek Visitor Centre won the Design Institute of Australia, South Australia Award, in the same year.

[30] Tim Shannon in an address to the staff of HASSELL in Sydney, 29th October, 2002.
[31] Christopher Wren in discussion with the writer, 7th November 2002.

For planning and landscape, the practice has won numerous accolades including the Australian Institute of Landscape Architects (AILA) National Project Award in Landscape Architecture in 2000, for Design, Rehabilitation and Conservation of the Millennium Parklands and they capped this with a second award in the Master Planning category—again for the Millennium Parklands. In 2002 the practice won the AILA National Project Award in the Master Planning Category for Kelvin Grove Urban Village in Brisbane.

Recognition by their peers can be seen as a benchmark of HASSELL's standing in the profession. In 1998, reviewing 60 years work by the HASSELL organisation, Tim Shannon noted, "our legacies are distinctive architecture and outstanding places".[32]

Current projects, heading into construction, which display the firm's exceptional design skills include the Commonwealth Law Courts in Adelaide. Conceptualised by the same team that was responsible for the highly praised Commonwealth Law Courts in Melbourne, the new court building elbows into the south-east corner of Victoria Square, employing a similar extrovert language, intended to make the building more accessible and its business more transparent.

The Bendigo Performing Arts Centre in Victoria draws upon the expertise acquired by the practice in the design of some 28 theatres including the National Institute of Dramatic Art in Sydney and the Roma Mitchell Arts Centre in Adelaide. The central 'drum' of the Bendigo auditorium is separated from the 'box' that houses the foyer by a strip of natural daylight, dramatically illuminating the interior and the vertical circulation. The foyer itself is extensively glazed so that, like NIDA, the theatregoers become part of the 'performance'.

The 16-storey Doncaster Hills project in Victoria incorporates current thinking on sustainability and mixed-use development. Twelve floors of apartments are stacked above four floors of retail and commercial units. Most of the 150 apartments have winter-garden terraces or balconies, and the complex cross-section, which takes some precedents from the Newington Housing at Olympic Park, results in highly modelled façades, which are accentuated by the discriminating use of primary-coloured panels. Other projects, such as the Parramatta Rail Link and the Liverpool-Parramatta Transitway Stations in NSW, build upon HASSELL's expertise in the design of transport infrastructure. When complete these projects will substantially redefine the image of Australian public transport.

Into Asia

At the beginning of the 21st Century, HASSELL is operating in the international arena.[33] The practice, encompassing architecture, urban design, planning, landscape architecture and interior design, has grown from a small architectural firm in Adelaide to be a major design collaborative working throughout the Asia-Pacific region, and with a strong presence in the Peoples Republic of China .

Building upon their expertise in the design of transport infrastructure HASSELL is currently working on four underground stations on the

---

[32] Tim Shannon in HASSELL 60th Anniversary publication, November 1998.
[33] World Architecture magazine in its January 2003 survey of international architectural practices reported that HASSELL are placed 61st in the top 500 practices in the world.

Above (both images): MRT Stations (project) Singapore.

Opposite: Ningbo Urban Design Master Plan Ningbo, Peoples Republic of China.

Singapore Mass Rapid Transit System (MRT) Circle Line extension.

The robust palette of materials is detailed to the same rigorous standards as the Olympic Park Rail Station and the Qantas Domestic Terminal in Sydney, and natural daylight is introduced into the station interiors through a unique roof and ceiling concept. The stations incorporate the clarity of circulation and smooth passenger flow that has become a recognisable feature of HASSELL's transportation projects in Australia. The Singapore MRT has acquired an international reputation for its efficiency and quality of design, and they have sought out a number of world-renowned design practices for the current generation of MRT stations.

In March 2003, HASSELL was successful in a limited competition for the Urban Design Master Plan of the Chinese city of Ningbo. As I noted in my introductory remarks on 'The Grid and the Landscape', many of HASSELL's projects are located at the intersection of the city planning grid and open space. In a manner that is reminiscent of Colonel William Light's plan for Adelaide, the Ningbo plan includes a number of urban squares and linear parks within a formal grid. The orthogonal form of the plan and its north south orientation, incorporating positive *feng shui*, also stirs memories of one of the two distinct models of ancient Chinese cities—Chang'an, the Tang dynasty city built from the 7th to the 9th centuries.

But comparisons with Chang'an can only go so far in explaining the appeal of the Ningbo Plan to the adjudicators. It has other layers of meaning. Superimposed upon the road infrastructure and the street blocks is a grid of waterways and canals, which provokes comparison with Asian aquatic cities such as Suzhou and Ayuthaya (the latter also employed an orthogonal pattern when construction commenced in 1351 for King Uthong, a Thai ruler of Chinese origin).

Unlike these historical precedents Ningbo will be a high-rise, high-density city, and in this sense the design recognises the positive values of the grid as an organisational device in recent high-rise cities around the world.

Ken Maher notes that these successes in the highly competitive markets of Australia and Asia indicate that "The future will be built upon our best work. Work which is intelligent and responsive to its setting and the aspirations of its people."[34]

34 Ken Maher in discussion with the writer March 2003.

HEIDELBERG / VICTORIA / AUSTRALIA / 1992–1994

Northern Metropolitan Institute of TAFE

In suburban Braybrook, alongside the highway from Melbourne to Ballarat, a beautiful but near derelict factory designed in the 1940s by Frederick Romberg stands in sharp contrast to the overall clutter. Its refined structure is a sharp appetiser for the Metropolitan TAFE building at Heidelberg, designed by HASSELL.

The Northern Metropolitan Institute of TAFE is, on first acquaintance, an austere building, hard-edged and unromantic with an industrial aesthetic, not inappropriate for a Technical and Further Education (TAFE) College. The building is tough and uncompromising, qualities also found in Australian vernacular architecture.

The design is strongly reminiscent of the "most famous Brutalist building" in England, namely Hunstanton School, Norfolk, by Peter and Alison Smithson (1949-54), which was an "uncompromising exhibition of materials".[1] One of the first great post-war schools in Britain, Hunstanton School used "exposed steel frame and brick wall panels of a Miesian background… it was hard almost to the point of brittleness".[2] There are also strong memories of the Illinois Institute of Technology (1940) by Mies van der Rohe in the design of the TAFE buildings.

"Brutalism implied some form of attempt to make manifest the moral imperatives that were built into the tradition of modern architecture by the pioneers of the 19th century".[3] These are descriptions that could equally be applied to the TAFE College.

The TAFE building also has an exposed steel frame with infill panels of off-white masonry and large aluminium framed windows. The aim was to achieve a high degree of transparency – admitting natural light for the students and showcasing the technology. Steel fire escape staircases are boldly expressed externally.

The campus was developed as an Advanced Manufacturing Engineering and Building Industries Training Centre for manufacturing, hi-tech industries, electrical, electronics and building construction programmes. The planning of the project is based on a central pedestrian precinct with a perimeter road providing vehicular access. The centrepiece of the design is the Advanced Engineering facility which has been used as a model for similar facilities throughout Australia.

This project has a marked affinity with the architectural language of an earlier HASSELL building, the CSR Bacchus Marsh Offices and Workshop, designed by Jack McConnell and completed in 1960.

Opposite and Above: The exposed steel frame and large aluminium framed windows achieve a high degree of transparency.

[1] Peter Murray & Stephen Trembly, writing in *Modern British Architecture since 1945*, Frederick Muller Limited, London, 1984.
[2] G. E. Kidder-Smith, in *The New European Architecture*, Pelican Books, London, 1961.
[3] Wolfgang Pehnt (Ed), in *Encyclopaedia of Modern Architecture*, World of Art Library, Thames and Hudson, London, 1963.

Opposite and Below: The Northern Metropolitan Institute of TAFE is an austere building, hard-edged and unromantic with an industrial aesthetic, not inappropriate for a TAFE Institute.

ADELAIDE / SOUTH AUSTRALIA / AUSTRALIA / 1996–1997

Bankers Trust Service Centre

The Australian Bankers Trust Service Centre at Science Park, in the southern suburbs of Adelaide, functions as a high-tech customer service and processing centre for Bankers Trust back-office operations.

The two-storey building is an energy efficient and paperless office designed with back-up power generation for continuous 24-hour operations. The facility combines flexible design and interior planning with sophisticated information technology and data systems.

Right: Linear office wings defines a series of well-scaled courtyard spaces.

section

BENALLA / VICTORIA / AUSTRALIA / 1993–1995

ADI Ammunition Manufacturing Complex

Set in a rural wetland, this modest collection of 60 functional buildings provides a sense of order and rigour that is both unexpected and refreshing. The project involved the construction of an integrated ammunition manufacturing facility for the Australian Defence Industries (ADI) on the outskirts of Benalla. Operations at the facility are the manufacture and assembly of ammunition, ranging from 5.5mm small arms ammunition through to 155mm artillery shells, and include the forging, machining, heat treatment and surface finishing of metal and synthetic products.

The architectural language is tough and unromantic, and establishes a connection with the early International Modernism of HASSELL's first generation. It is also a clear companion to the industrial aesthetic of the Northern Metropolitan Institute of TAFE at Heidelberg, completed in 1994.

Right: Steel portal frames provide an ordering to the main façades.

Below: Masonry end walls are simply modelled and relate to the landscape setting.

WERRIBEE / VICTORIA / AUSTRALIA / 1993–1997

Victoria's Open Range Zoo

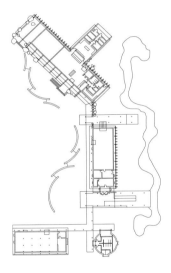

plan

In 1992 HASSELL prepared a masterplan setting out the conceptual framework of this world class open range zoo on a 178 hectare site, 31 km from the Melbourne CBD. The zoo replicates the savannah habitats of Africa, Australia and Asia, and it presents an environment where grasslands are conserved, and their importance as ecosystems and survival habitats are explained.

The zoo has a core visitor's area linked to a series of themed areas by walking trails or safari bus. The architecture of Victoria's Open Range Zoo at Werribee indicates simplicity of form and richness of material and detail. The buildings, consisting of an interpretative centre, visitor facilities, restaurant and exhibition pavilions, are steel framed and clad in timber and galvanised steel, contrasting with the rammed earth walls which harmonize with the savannah theme.

The Zoo has a low impact on the environment as it produces most of its own animal foods, it recycles most of its own wastes and it operates a low energy regime. It also contributes to the conservation and reproduction of endangered species of animals.

Above: Delicate shade canopies provide mediation between the visitor's centre and the park landscape.
Opposite: Pavilions with directly expressed frames and horizontal proportions define the edge of the open plain setting.

ADELAIDE / SOUTH AUSTRALIA / AUSTRALIA / 1975–1997

River Torrens Linear Park

HASSELL were among the first practices in Australia to adopt an ecologically sensitive approach to architecture and planning, involving landscape planners at a very early stage in projects.

In 1975 the firm was commissioned to draw up a master plan for the River Torrens Linear Park and Flood Mitigation scheme, a 32 kilometre linear park stretching from the Adelaide Hills to St. Vincents Gulf. The firm's involvement lasted more than 20 years and most recently has involved the design of the Adelaide Riverbank Promenade.

A pedestrian and cyclist trail has been provided along the entire length of the river, and the many diverse environments and land uses have been integrated by the planting of a dominant canopy tree, the River Red Gum *(Eucalyptus camaldulensis)*. Areas of historic conservation have been incorporated, and a continuous wildlife corridor has been created through 30 kilometres of intensively developed urban area.

In 1995 the practice was engaged for an expanded study and the preparation of a Comprehensive Catchment Water Management Plan for the whole area.

Right: A series of landscape settings reinforces the natural features of the park, and provides an ordered sequence to the park journey.

MELBOURNE / VICTORIA / AUSTRALIA / 1991–1995

St Vincent's Public Hospital

Melbourne's 400-plus-bed St Vincent's Public Hospital is considered one of the nation's leading public teaching hospitals. The architecture and interior design provided by HASSELL has set a new level of quality and ambience for public healthcare institutions in Australia, and introduced the "patient focussed care model" that is reflected throughout all the hospital's services. As a model for subsequent private and public hospital design, it has generated a high degree of international interest.

Right: The spatial quality and materiality of the main foyer sets the tone for the whole complex.

SYDNEY / NEW SOUTH WALES / AUSTRALIA / 1995–1997

Chifley Square and Café

Left: A public café situated within a predominantly corporate space is both eloquent and reserved yet approachable.

Below: The strong spatial composition of the square is a direct response to the curious geometry of a former tram terminus on the site.

Chifley Square is named after Ben Chifley (1885-1951), an engine driver, unionist and socialist politician from Bathurst who became Australia's WW II treasurer and post-war Labor Prime Minister. An eight metre high steel-plate statue of Chifley (with his familiar tobacco pipe in hand and crumpled suit), fabricated by artist Simeon Nelson, dominates the square.[1]

Appropriately, HASSELL's urban design scheme for Chifley Square can be seen as democratic, with the insertion of a public facility - a café and outdoor terrace – into an area of predominantly corporate culture.

The square is D shaped, a curious geometry that came about in 1937 because it was a terminus for Sydney's tram system. HASSELL have used this curious shape as the starting point for their urban intervention, which is a simple yet strong spatial composition.

One edge of Chifley Square is defined by the 46 metre high, sensuously curved green-glass façade of Qantas House (1957) designed by Rudder Littlemore and Rudder. The curve is picked up by the Wentworth Hotel, designed by Skidmore Owings and Merrill, and by the 41-storey Chifley Tower, designed by Kohn Pederson Fox (USA) and Travis Partners.

The 120 square metre café is a linear sculptural composition of grey granite and glass, with a folded plane zinc-clad roof. A semi-opaque glass wall, which forms the back of the restaurant, defines the footpath line on Hunter Street. The 1.2 metre thick 'wall' accommodates the kitchen, toilets and the furniture store. Thus anchored, the café opens out to the northern open space to capture the sun.

The wall has been extruded and projected to the west as a prismatic sculptural object, termed the Lightwall Crucimatralux installation, again designed by Simeon Nelson (in this case a collaboration with HASSELL). Semi-opaque glass gives way to transparent glass fixed to a milled finish stainless steel frame. Within it are overlapping diagonal sheets of glass, and there is a play of refracted light moving into infinity.[2]

The main axis of the urban project is east-west. Linear benches of recycled hardwood with protruding stainless steel knobs (to foil Sydney's ubiquitous skate-boarders), and bluestone paving inlaid with strips of Verde granite, follow this geometry, as does the sculpture of Chifley, which is formed by two 20 mm-thick cut-outs of stainless steel fixed to a steel frame. The form of the statue and the materials are an unequivocal comment on the many "tortured bronze likenesses of Queen Victoria and explorers" that abound in Sydney.[3]

Tall cabbage-palm trees are planted in a linear pattern that intersects with the geometry of the cafe at an acute angle. To maintain a consistent grid, Philip Street was realigned to accommodate a row of cabbage-palms in front of Qantas House with another row in the median strip. The trees have the effect of unifying the entire square.

The design of the square is a sophisticated understatement that is probably more befitting of Ben Chifley, "an uncommon common man", than a ceremonial space.[4] One writer has noted that the cafe is "like a handsome interpretation of the Barcelona Pavilion relocated to Australia".[5]

1 Elizabeth Farrelly, 'Chifley gets a nip and tuck', writing in *The Sydney Morning Herald*, Sydney, October 28th. 1997.
2 Peter Ward, 'Sydney's 12-point plan', in *The Weekend Australian*, December 13th–14th, 1997.
3 Peter Ward, ibid.
4 Peter Ward, ibid.
5 Brian Zulaikha, 'Chifley Square', writing in *Architecture Australia*, March/April, 1998.

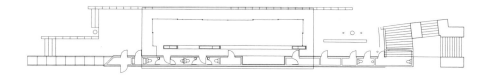
plan

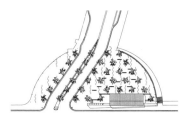
site plan

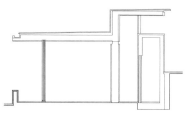
section

Left: A semi-opaque glass wall forming the back of the café defines the footpath line on Hunter Street.

Below: Tall cabbage-palm trees are planted in a linear pattern, unifying the entire square.

HONG KONG / PEOPLES REPUBLIC OF CHINA / 1992–1997

Tai Po Waterfront Park

Tai Po, on Tolo Harbour 20km north of the Hong Kong city centre, is one of Hong Kong's original industrial areas. Tai Po Waterfront Park, designed for the Hong Kong Government, is a 15-hectare recreation venue. HASSELL was appointed as the lead consultant for the project, which was used as the site for the sovereignty handover ceremony in July 1997.

The dramatic silhouette of the lookout tower is now a symbol of the Tai Po area, and has become a tourist attraction in its own right. The park itself has been extraordinarily successful with the local residents: school children, *tai chi* practitioners and courting couples thronging the park throughout the day and the evening. The park has a strong axial form reinforced by the innovative use of materials including softwood timber, concrete, steel, glass and water.

The design questions preconceptions of what an urban park is, and what it should be. The formal landscaping and the spectacular floral displays are lovingly maintained, and Tai Po Waterfront Park is an ongoing success story born out of a society in transition and focusing on the future. The park represents the integration of architecture and landscape reflecting the unique status of Hong Kong in 1997 – an international city undergoing momentous change.

Opposite and right: The tapering form of the viewing tower is created by a spiral walkway built from laminated timber. A complex geometry is displayed in the steel structure supporting the walkway.

Left and below: Walled formal gardens are lovingly maintained, and create a spiritually soothing environment as a refuge from the town of Tai Po. Water is used sparingly, but in an architectural manner to reinforce the axial form of the park.

Overleaf: The tapering form of the viewing tower is surrounded with a circular moat and by sloping beds of exotic plants from around the world.

site plan

MELBOURNE / VICTORIA / AUSTRALIA / 1997–

Minter Ellison Offices

The client floor for Minter Ellison Lawyers is located in the north tower of the Rialto Building in Collins Street, Melbourne, extending the existing meeting facilities to the whole level. The design expresses the core values and aspirations of Minter Ellison, communicating a consistent corporate statement to their clients.

A diverse range of meeting settings employ varying degrees of formality, scale, and usage to create five individually expressed elements.

The formal, large scale common conference rooms and the large scale dining room use geometry as a strong motif. The *red light* rooms offer a medium scale, semi-formal intimate environment. In contrast, the informal *blue light* rooms create an open focus. As support, a series of curved glass *'pod'* rooms offer informal, small-scale interview spaces.

Right: Corner meeting room dominated by an illuminated red fabric covered cylinder.

Left: The partners' lounge is backed by a wall of slipped-form joinery, concealing doors and providing shelving.

TOOWOOMBA / QUEENSLAND / AUSTRALIA / 1997–1998

Empire Theatre

The Empire Theatre is the largest regional theatre in Australia. Originally built in 1911, and then rebuilt in 1933 as an Art Deco cinema, the theatre is a listed heritage building with a long history as the social and cultural hub of Toowoomba. The theatre closed in 1971 and fell into disrepair.

The Toowoomba City Council commissioned HASSELL, in June 1995, to refurbish the former cinema as a 1573 seat performing arts centre. Approximately 3,500 square metre of existing heritage-listed space was refurbished, with 4,000 square metres of new space added to the building. The heritage auditorium, foyers and existing facades have all been meticulously restored. The 1911 brick side walls have been retained in their original condition in the side foyers, and the 1933 riveted steel girder structure is revealed.

The new work, which includes new side foyers, bar facilities, stage, fly tower and rehearsal rooms, has been carried out as part of the redevelopment of the theatre and is intended to evoke the spirit of the 1930s in a contemporary manner. The new additions touch the existing building fabric lightly in order to clearly distinguish between the new and the old.

Right: The original Art Deco interiors of the foyers have been meticulously restored.

Opposite: Two new side foyers provide a contemporary interpretation of the original highly decorated façades.

Left: The recessive nature of the additions allows the historic central building to dominate the composition.

Below: The unusually large auditorium with distinctive Art Deco lighting and decoration has been meticulously restored.

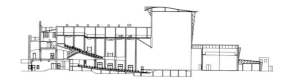

section

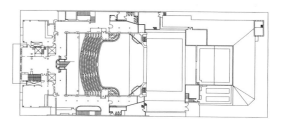

plan

UNIVERSITY OF SYDNEY / NEW SOUTH WALES / AUSTRALIA / 1997–1998

Gordon Yu-Hoi Chiu Building

The Gordon Yu-Hoi Chiu Building is a private research institution on the Sydney University campus. Funded by a former alumnus, the new building is a triangular shaped annexe to an existing three storey Chemical Engineering Building, designed by Ancher Mortlock and Woolley.

HASSELL worked within an existing masterplan, which defined the general building form. Their strategy has been to relate the prismatic geometry of the new building to the existing building, but to articulate the new structure clearly through physical separation and a distinctive architectural language. Between the two buildings is an external wet laboratory area of linear configuration, covered by a glass roof and separated by a 'blade' wall. The new building connects to the old via an open bridge, which spans the intervening gap and dramatises the movement from old to new.

The elegant steel and glass extension celebrates the view to the north towards Maze Green. The building is slightly elevated above this landscaped open space at the heart of the campus, surrounded by buildings of brick and concrete construction from an earlier era. To the west there is an entrance court and a pedestrian route into the University. The west wall of the Yu-Hoi Chiu Building responds to this, and is designed as a solid 'edge' which defines the court. Professorial suites in the research facility penetrate this wall and overlook the court.

The new research facility contains flexible double height laboratory spaces on the ground floor overlooked by a mezzanine level with offices. The architects' intention was to produce clarity of form and space, and to reduce the complexity of the elements to a minimum. The interior employs interlocking geometry and is dramatised by the penetration of daylight.

The building is of steel frame construction, with Vitrethane panel cladding used externally. The detailing expresses lightness and transparency. Façade mullions are 125 mm x 125 mm galvanised steel flat sections with aluminium cover plates. These mullions also provide structural support for the roof.

The bulk of the low-key research building is not extravagant, and is constructed from utilitarian materials such as galvanised steel, painted Villaboard and cement floors. The more costly details and custom finishes such as aluminium and stainless steel are employed on the northern façade, which is the public front of the building.

Opposite: The detailing of the elegant steel and glass façade expresses lightness and transparency.

Left: The northern façade looks over Maze Green, providing a solid 'edge' to a large open space on the university campus.

Below: The triangular geometry of the plan is shown by the penetration of the upper-level office floor through the perimeter blade wall.

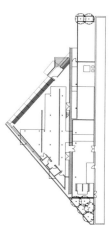

plan

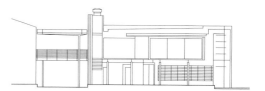

section

SYDNEY OLYMPIC PARK / NEW SOUTH WALES / AUSTRALIA / 1996–1998

Olympic Park Rail Station

Commissioned in 1995, and completed in 1998 for the 2000 Olympic Games, the Olympic Park Rail Station at Homebush is HASSELL's most widely recognised and acclaimed building. Eighty per cent of the 1.2 million people who attended the 2000 Olympics arrived by train, and the station deals comfortably with passenger flows of up to 50,000 people an hour at peak periods.

As with Nicholas Grimshaw and Partners design for Waterloo International Terminal in London (1993), the Olympic Park Rail Station is in the tradition of the great glass and iron railway stations of the 19th Century, such as London's St Pancras and the Gare du Nord in Paris, both built in the 1860s.

The planning, structure and materials are all directed towards achieving a smooth flow and rapid turnaround of passengers at major events. The single rail loop from Sydney's western line is mostly above ground, but it dives below ground for 1 kilometre before arriving at the station, and re-emerges as it exits on the other side. Arrivals alight on the central platform and ascend to ground level before surging into the square, while departing passengers board from the two side platforms.

The power of the design, by Rodney Uren and Ken Maher, is in its conceptual clarity and legibility for both arriving and departing passengers. Uren was able to draw on experience gained from working in Norman Foster's office on the Bilbao mass transit system and the Canary Wharf Station in London, while Maher introduced a strong landscape intrepretation.

The station is cut into the ground, with a simple dual element composition – one of base and canopy – and an unambiguous sense of direction. The escalators all run in the same direction, and the direction of egress is instantly recognisable. Glass lifts are provided for disabled access, thereby fulfilling a key requirement of the Olympic Co-ordination Authority (OCA) that there should be equal access for all. The lifts accommodate up to four wheel chairs. Tactile paving and signage is designed to assist the visually impaired, and induction loops are incorporated to assist those with hearing aids.

Essentially the building has two separate elements; the concrete substructure incorporates a central platform and two side platforms and has a strong and durable character: the elegant vaulted roof gives the building an immediate visual presence and provides shelter. Throughout there is a restrained use of colour, and the two-storey concourse lined with off-white precast panels has a monumental quality befitting a great public building.

The roof was conceptualised as a 'leaf like' structure, deriving inspiration from the native gum tree. It is a 200 metre open-ended canopy with 18 arched Vierendel steel trusses, stretching like a giant silver chrysalis from east to west in the landscape. The diaphanous steel roof structure that soars over the passengers springs from 11 metre high precast concrete columns that rise from the platform level. Externally the roof is clad in the quintessential Australian building material, profiled zincalume-coated steel decking, with glazed strips along the ridge of each vault admitting daylight. The linear emphasis of the station's form sets it apart from the Olympic Park sports venues.

There are no services in the roof. Lighting is directed upwards from below onto the white perforated

Opposite: The roof was conceptualised as a 'leaf-like' structure, deriving inspiration from the native gum tree.

Left: The vaulted entrance canopy celebrates a 'sense of arrival' in the tradition of all great railway stations.

Overleaf: The 200 metre canopy stretches like a giant chrysalis from east to west in the landscape.

aluminium acoustic ceiling panels. The impression from inside the station is that the roof is floating lightly above the base. Externally it appears more embedded in the land.

The rail station is designed to be environmentally friendly, constructed with materials that have a low level of embodied energy. The structure acts as a self-cooling breezeway, which also utilises natural light. Recycled timber is used for the cladding of station management offices and for handrails.

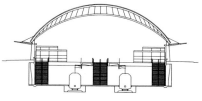

section

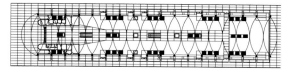

plan

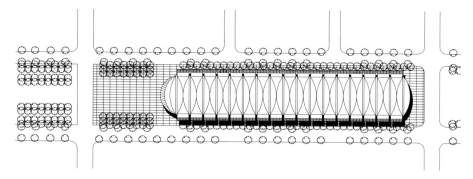

site plan

Left: The diaphanous steel roof structure, springing from 11 metre high precast concrete columns, soars over departing passengers.

Below: A strong presence for the station within Sydney's Olympic Park is achieved through horizontal rather than vertical scale.

Below: The roof is clad in the quintessential Australian building material, profiled zincalume-coated steel decking, with glazed strips along the ridge of each vault admitting daylight.

Opposite: The structure is carefully crafted, and the design concept has been rigorously followed through in the detailing.

CANBERRA / AUSTRALIAN CAPITAL TERRITORY / AUSTRALIA / 1998–1999

Canberra Playhouse Theatre

Opposite and Above: A transparent rectangular box containing the foyer contrasts with the cylindrical mass of the auditorium.

The Canberra Playhouse Theatre is a 600-seat live performance venue, forming part of the cultural facilities in the Canberra Civic Square. It is located adjacent to the City Hill, a prominent feature in the Walter Burley Griffin City Plan. At an urban level, the Playhouse was required to enhance the formal composition of the Civic Square and simultaneously to have a presence on Vernon Circle.

HASSELL's team, led by Ken Maher with key input from John Morphett, resolved this requirement for duality in the plan form by using the cylindrical form of the auditorium as a fulcrum. Lobby spaces and support facilities are configured in a building mass which echoes the form of the nearby Canberra Theatre, and which relates to the axial composition of the Civic Square. The auditorium itself, together with the fly tower and the back stage accommodation, are rotated through 15 degrees resulting in the long axis being perpendicular to the radius of City Hill.

The published texts of British theatre designer, Iain MacIntosh were influential on the design. MacIntosh argues that theatres in the 20th Century have moved away from traditional theatre, and in the process have lost their intimacy. He advocates a return to the Shakespearean form with balconies and a reduced distance to the stage. so that there is greater engagement between the audience and the actors.

Consequently the Canberra Playhouse auditorium takes a semicircular form and is designed with three levels including stalls, upper and lower balconies and several private boxes. The stalls seating has curved rows focusing attention on the centre of the stage. The maximum distance from the front edge of the forestage to the back row is 15 metres. Both balconies have three rows of seats to the rear of the auditorium and two rows at the sides. The maximum viewing angle at the centre of the second balcony is approximately 27 degrees below horizontal. The interior of the theatre is lined with timber, utilising a three dimensional pattern of panels to ensure effective acoustics.

The foyer space was conceived as a transparent rectangular box enveloping the solid form of the cylindrical auditorium. The foyer looks out over the Civic Square, and although the foyer space and bar facilities appear to be tight, this is partly compensated by spillover space in the adjoining link building between the Playhouse and The Canberra Theatre.

Despite these spatial limitations and a limited budget, the Playhouse has been acclaimed as one of Australia's leading theatrical venues and has invigorated the cultural life of the Federal Capital. It reflects a late 20th Century return to the meaning of theatre not merely as a place to see and be seen, but as a place to seriously engage with the actors and the performance.

Left: The theatre mediates with the formal plan of Walter Burley Griffin and the axial composition of Civic Square.
Below: The semi-circular form of the auditorium provides strong visual links between actor and audience.

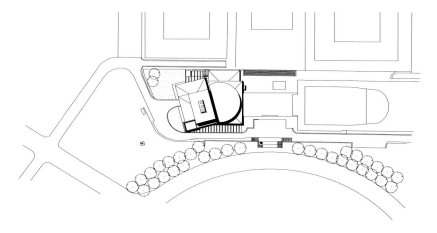

site plan

HONG KONG / PEOPLES REPUBLIC OF CHINA / 1997–1999

Ohel Leah Synagogue

The Ohel Leah Synagogue was built in 1901 to serve the Jewish community in Hong Kong. HASSELL collaborated with conservation specialists from the South Australia Government Architects Division to carry out a comprehensive conservation exercise intended to bring the building and grounds up to the current needs of the Jewish community.

The work included removal and replacement of roofing material and external render, stabilisation of the original building fabric and new damp proof measures. Lead lights were replaced, and appropriate furniture and fittings designed.

A significant aspect of this project is the design of the urban spaces surrounding the historic building. This work involved significant new architectural elements dealing with address and entry. An uncompromisingly Modernist position has been taken with these new steel and glass elements, proving a distinctive contrast to the massive masonry character of the original building. At night, the new work acts as a beacon for the Synagogue, signalling its presence in the city.

Opposite: The Modernist design of the new entry spaces contrasts with the 100 year old synagogue.

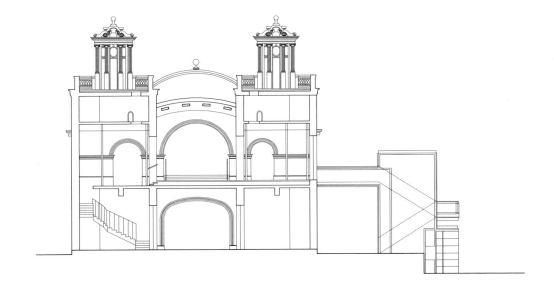

section

Opposite: The sensitive restoration of the synagogue affirms the presence of the Jewish community in Hong Kong.

Below: The new steel and glass elements create a formal entry to the synagogue.

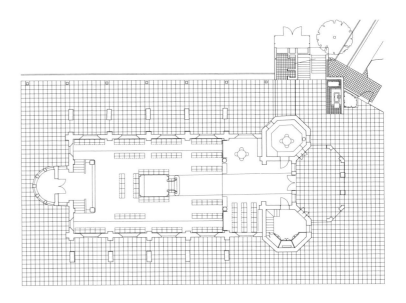

plan

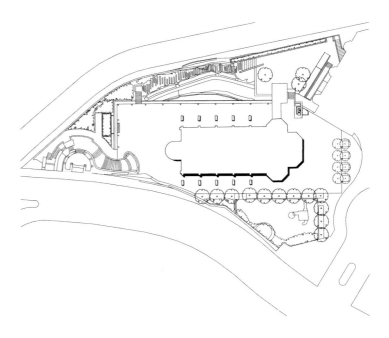

site plan

SYDNEY / NEW SOUTH WALES / AUSTRALIA / 1997–1999

Millennium Parklands

Above: The key elements which structure the landscape are: lowlands; walls and rooms; and elevated landforms.

Opposite: A significant wetland system dominates the previously contaminated industrial site, providing a complex system of water storage to harvest runoff from Newington (the former Olympic Village) and adjacent sites.

The Sydney 2000 Olympic Games focused considerable international attention on the Millennium Parklands and the Homebush Bay site.

HASSELL (in association with Peter Walker & Partners and Bruce MacKenzie Design) was selected by the Olympic Co-ordination Authority to prepare the concept plan for the 500-hectare Millennium Park, an area one and a half times bigger than Central Park in New York. The resultant plan has created a new model for the world's metropolitan parklands in the 21st Century.

The purpose of the concept plan was to bring together the diverse landscapes, memories and cultural elements within the parklands. The HASSELL team, led by Tony McCormick, drew upon a wide range of specialist skills. There was a cross-fertilisation of ideas between landscape architects and architects to develop the wetland system and the landforms. The key design elements in the physical plan were:

Lowlands:
The development of the Haslams Creek lowlands into a park core featuring open and forested wetlands of salt and fresh water. These together with the lowlands of Homebush Bay, Bicentennial Park, Boundary and Powell Creeks provide a lacework of riverine and estuarine habitats that penetrate the Homebush site.

Walls and Rooms:
A linear system of woodlands and forests *(walls)* defining spaces and settings *(rooms)*, through which roads, pedestrians and cycle trails connect. In addition to spatial articulation, the *walls* of varying widths play a key role in providing habitat diversity.

Elevated landforms:
A series of constructed elevated naturalistic and geometric landforms that are accessible by bike, wheelchair and foot. These landforms provide identification, orientation and understanding of the complexity of the low-relief landscape. They are also used for burying contaminated material.

The plan demonstrates 'best practice' principles at many levels including park planning, design and programmes for education conservation and recreation. A network of 40 km of pedestrian and cycle trails are provided and over 2.5 million plants will eventually be planted in the completed parkland. Habitats have been created for the Latham's Snipe and the endangered Green and Golden Bell Frog species, as well as 10 migratory species of bird.

The Hill Road Corridor is the first application of the principles of the Millennium Parklands Concept Plan. It incorporates all the features of the 'lowlands', 'walls' and 'elevated landforms' outlined above. The implementation was based upon ecologically sustainable processes. Stormwater collection, treatment and reuse were the central driving forces generating the design, with a complex system of water storage to harvest the runoff from the Olympic Village and the wider watershed. Landforms at the northern end of the corridor contain 1.5 million cubic metres of contaminated fill from the Homebush Bay site. A topsoil 'facsimile' was reconstructed for use in the corridor from imported sandstone fill and recycled organic material.

Top: The Haslams Creek lowlands take on an informal character and provide estuarine habitats.

Above: Water storage ponds are expressed as geometric forms, revealing the constructed nature of the landscape.

Opposite: The Concept Plan clearly shows the provision of a diverse landscape with strongly modelled landforms.

SYDNEY / NEW SOUTH WALES / AUSTRALIA / 1999

Qantas Domestic Terminal

Opposite: The bold formal approach of the architecture is clearly expressed, both externally and internally.

HASSELL has a special relationship with Qantas. The practice was responsible for the refurbishment of the International Terminal at Melbourne Airport and is currently working on the Adelaide Airport. They have acquired a special expertise in this building type and are now one of the foremost designers of transportation interchanges in the Asia Pacific region.

In the late 1980s HASSELL won a limited competition for the Qantas Domestic Terminal at Sydney Airport, with a bold yet simple formal solution with a strong spatial quality. The project was deferred for some time, and was reactivated in 1995 with a design team led by Tony Grist. The completed building has become the benchmark solution in terms of design quality, materials and accommodation for Qantas throughout Australia, and represents the first international quality terminal in the company's domestic network.

The terminal is a lofty light-filled volume. Its footprint was strictly constrained by existing aircraft taxiways and parking bays on the apron, but it is designed as a tiered building with a lightweight structure on a heavy weight horizontal plinth. The wide-span roof structure is a gently curved truss, supported on quadruped pillars, with linear roof lights providing an even quality of daylight throughout the departures concourse. The departure concourse terminates in a lofty circular hall serving six gates.

The geometry of the roof lights supports the design intention to give directional clues. The aim was to provide visual clarity and make the routes for both departing and arriving passengers instantly legible. Clarity and legibility are recurring themes in HASSELL's architecture (see also the Sydney Olympic Park Rail Station).

The intention was that the passengers should also have a visual connection with the aircraft on the apron, and this is achieved by having a transparent skin at all levels. This immediately assists with way-finding and involves passengers in the activities of the airport. Passengers in the departure concourse are in such close proximity to the tarmac that they can observe the aircraft crew carrying out pre-flight checks and the ground crews servicing the aircraft.

Retail outlets are arranged so that they do not cut off the view of the activities on the tarmac. The aim was to design an airport with retail facilities rather than a shopping mall with peripheral airport activities.

Similar frameless glass at the lower arrivals concourse allows the internal space to flow out quickly and efficiently to the external space, and for meeters and greeters to be immediately identifiable and accessible to arriving passengers.

Many of the design decisions were motivated by pragmatic concerns, such as the need for a column-free space in the baggage handling area, which resulted in a pre-stressed slab. The structural grid is 9.6 metres, determined by the aircraft parking grid. The weight of the departures hall roof was reduced significantly by the use of semi-prefabrication, and three-dimensional bracing was employed to reduce the spans.

There are no services in the concourse roof, which also contributes to its lightweight appearance. All lighting is from service 'trees', which also house the public address system and fire detection system. Lighting is directed upwards and reflected off the underside of the ceiling. The level of natural daylight in the terminal helps reduce the artificial lighting

requirements. The building employs principles of fire engineering rather than conventional fire escape criteria. Local air-conditioners employing jet diffusers are used around the habitable spaces and each check-in counter has dedicated air-conditioning.

Call time for Qantas domestic flights is 15 minutes, and turn around time for aircraft can be cut to 30 minutes. This turn around time is crucial, as it is the only way to run a domestic airline economically. An aircraft that is not in the air is not earning money, so design innovations that cut the time on the ground are critical.

The terminal building captures the exhilarating feeling of flight. The departures hall is a grand space in the tradition of transport interchanges around the world, and gives a sense of the high technology involved in 21st Century communications. There is an atmosphere of unhurried calm, sufficient to soothe even the most anxious passenger.

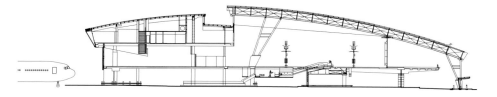

section

Left: A calm and soothing ambience is the result of the muted and seductive use of colour, and the fluent rhythm of the roof structure.

Left: The wide-span roof structure of gently curved trusses is supported on quadruped pillars.

Above: Linear skylights provide an even quality of daylight throughout the departures concourse.

MELBOURNE / VICTORIA / AUSTRALIA / 1995–1998

Commonwealth Law Courts

Melbourne's Commonwealth Law Court complex is Australia's most important legal building outside the national capital. The brief gave Tim Shannon and Paul Katsieris of HASSELL's Melbourne office the opportunity to explore the notion of what a public building should look like at the outset of the 21st Century. Many functions that were formerly in the public realm are now privatised, and the traditional definition of public architecture has become blurred.[1] The Law Court complex, Katsieris suggests, is one of the last truly public buildings.

Located at the corner of La Trobe and William Street, the building overlooks a public space facing the Flagstaff Gardens. The building occupies a whole city block measuring approximately 100 metres x 100 metres and consolidates the High Court, the Federal Court and the Family Court, a total of 43 courtrooms and hearing rooms, into a 17 storey building (including basement) consisting of two blocks, separated by a dramatic north-south galleria and linked by footbridges. The rear block, a simple rectangle, is angled at three degrees away from a 'T' shaped block that faces the William Street entrance. This slight entasis gives a funnelling effect that enhances the drama of the internal street. Katsieris suggests that the true meaning of entasis is that it is a device that "exhibits tension" and is used to "make things look alive".

Katsieris explains that the design of the Law Courts " brings together many themes HASSELL have been working on for years". It was about "setting up order and then destabilising it" and "moving the eye around". Appropriately the building is concerned with order, which can only come about as a result of the tension between chaos and control. It is a building caught between movement and stillness.

Katsieris is a graduate of the University of Melbourne School of Architecture, where lectures by Miles Lewis and George Tibbits opened up many lines of enquiry. His major architectural influences have been Le Corbusier and Louis Kahn, and he acknowledges a fascination with ancient medieval works and with tapestry. Tim Shannon (who brought him into HASSELL) taught him in his final year. In 1988 and in 1990 he won a scholarship to study for two years in Greece

The 'Golden Mean' is used to compose the curtain walls of the Law Courts. Perhaps for this reason it stir memories of Le Corbusier's Salvation Army building in Paris (1931-32). Another building by Le Corbusier, the Secretariat in Chandigarh (1951-56) also comes to mind.

The lower floors of the Melbourne Law Court comprise courtrooms, hearing rooms and meeting rooms. The upper levels contain judges chambers, with the top floor devoted to the High Court. The horizontal layering of the judicial hierarchy possibly explains the fact that the design façades have a tripartite arrangement, which reverses the expectation that the bottom of the building should be heavy and the upper part light. Most senior judges are afforded a symbol of their status with an individual balcony. The functional zones are thus clearly expressed through the formal composition of the façades.

The building is clad in curtain walling, which employs several permutations of density and opacity within a standard grid. Katsieris adopts a strategy of misalignment, so that the underlying grid is not evident on the surface. But "like nuances in a jazz piece" the seemingly random ordering of the façade

Opposite: The syncopated façade of the Law Court complex engages with the grid of Melbourne's city centre and with the urban space of Flagstaff Gardens.

[1] Hamish Lyon, 'Hassell's Law', in *Architecture Australia*, Vol. 88, No. 5, September/October, 1999.

has an underlying rhythm.¹ Although balconies project in a random fashion and the seventh, eighth and ninth floors are recessed for no apparent reason, the whole 'hangs together' as a coherent and disciplined work of art.

Katsieris is also fascinated by the idea of a tapestry and, curiously, needlework, both of which can be seen as suitable metaphors for the façade fenestration patterns. He found inspiration for the copper, sea green and grey colours in the paintings of Giotto, while Mondrian's palette may have provided inspiration for the pivoted ceremonial entrance panel and forecourt pavilion - perhaps via Le Corbusier and the Chandigarh Secretariat building.

The entrance to the building is through a transparent box beneath a solid steel clad portal. Chapter 3 of the Constitution is engraved upon the glazed wall of the box, casting its shadow like a light veil over the judiciary and public alike. The main internal volume is a six-storey atrium extending the entire length of the city block. An exhilarating space, it is filled with light from elliptical skylights, punctuated by bridges and enlivened by a dynamic curved concrete stair.

Claimed to be the most technologically advanced Law Court building in Australia, the functionally complex building is handled with great clarity. The internal detailing and manipulation of surface texture befit a major public building which "demonstrates that architecture can still engender a level of idealism and optimism".² The building speaks of the character and dignity of the law, and simultaneously suggests an optimism about the future.

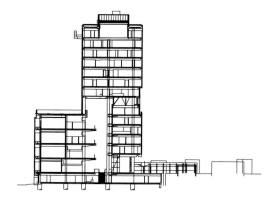

section

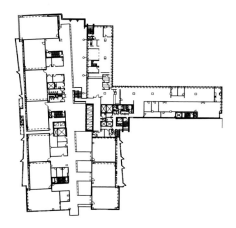

plan

2 Joe Rollo, *Contemporary Melbourne Architecture*, UNSW Press, Sydney, 1999.
3 Hamish Lyon, ibid.

Left: The façades have a tripartite arrangement, which reverses the expectation that the bottom of the building should be heavy and the upper part light.

Left, Below, Opposite and Overleaf: The six storey atrium is an exhilarating space, filled with light, punctuated by bridges and enlivened by a dynamic curved concrete stair.

Right: The curtain walling employs several permutations of density and opacity within a standard grid.

MELBOURNE / VICTORIA / AUSTRALIA / 1999

Helvetica Café

This project has two agendas, a social agenda and a marketing agenda. Helvetica embodies a movement towards unity within diversity with its multi-layered brief – licensed a la carte and function area, internal and external café areas, take away and fax orders, cake and coffee lounge, and a commercial kitchen supplying another three existing establishments – all within the realm of a public building.

The articulation and containment of these varying concerns occur within a framework that takes its cue from its built context. Consistent with the public spaces of the Commonwealth Law Courts building, the ground floor café, kitchen and coffee lounge seek an expression that is clearly organised, self-revealing, and full of activity. The mezzanine à la carte in turn shares the stance of the judicial chambers in the building above, offering a higher level of privacy and solidarity.

Helvetica's language of movement and shift within a tactile sense of modernity both reflects and expands on the dialogue offered by its site – light and views, clarity and activity, forward looking, continuity of form and materials from a macro level to a micro one.

Right: The linear form of the interior is enlivened by a coloured drum element.
Opposite: Artworks and patterned cladding provide a consistency between the café and the interior of the court complex.

MELBOURNE / VICTORIA / AUSTRALIA / 1996–2000

Alanbrae Residential Estate

Alanbrae Estate is divided into 400 residential lots. In addition to developing the master plan, HASSELL was responsible for the landscape design including a system of open spaces, a road bridge, a boardwalk and site furniture organised around a centrally located lake, which features a 'floating' pavilion.

The coordination of these elements at the outset has resulted in a high quality residential environment with a strong identity. Of particular interest is the delicate pavilion perched over the edge of the lake giving form and identity to the estate.

Right: The pavilion and lake provide a romantic diversion from the surrounding suburban housing.

SUNSHINE / VICTORIA / AUSTRALIA / 2000

Sunshine Courthouse and Police Station

Sunshine Courthouse and Police Station followed on the heels of the Northern Metropolitan TAFE at Heidelberg, and explores similar architectural ideas.

The Magistrates Court and Childrens Court is located in a challenging context, that is to say it is an area of commercial sprawl, a 'non-place' at the side of the railway tracks, with huge discount stores, giant advertising hoardings and extensive car parks.

In an environment where everything else is shouting for attention, the making of a civic building is difficult. HASSELL chose to react against this context with a mini-brutalist building; straightforward, hardy and frugal, without excess or decoration. The two-storey Courthouse building is clad in curtain walling with natural anodised aluminium mullions, set above a recessed sub basement that is partially clad in red brick.

The police station, which is entered at the rear of the Courthouse, is clad in the same red brick with a steel framed portico and pergola. In this manner the severe lines of the Courthouse and Police Station stand apart from their more consumerist orientated neighbours, upholding and representing the authority of the law. The curved entrance portico to the Court softens this otherwise stern building.

Top Left: The linear public gallery of the Courthouse.
Top Right: The Police Station with steel framed portico and pergola.
Below Left: The severe lines of the Magistrates Courthouse are softened by the curved entrance portico.
Below Right: The entry to the Childrens Courthouse.

SYDNEY / NEW SOUTH WALES / AUSTRALIA / 1997–1998

Fox Studios Car Park

Multi-storey car parks are the bane of city planners. A necessary evil, these large mundane structures are extremely difficult to integrate into the urban fabric. Internally they are invariably low ceilinged, gloomy, sparely lit, with grubby exit staircases. So it comes as pleasant surprise to drive into the Fox Studios Car Park on the site of the former Sydney Showgrounds at Moore Park.

The historic Showgrounds hosted the famous Royal Easter Show for 115 years, but following the Showgrounds' controversial relocation to the Olympic Park at Homebush Bay, the area was rezoned for filmmaking and leased to Fox Studios Australia. HASSELL was engaged as the co-ordinating architects and master planners for the Bent Street precinct, working with five design groups and with heritage consultants Godden Mackay Logan. In addition the firm were charged with the design of the public domain, including the Show Ring Park and a seven storey car park for 2000 vehicles.

The car park is one of the most successful buildings on the site. Employing a strictly functional Modern architectural language with an emphasis on tectonic clarity, it challenges the popular notion of international theme parks. It is a totally contemporary structure, which is nevertheless conscious of the heritage context and the scale of adjacent heritage buildings.

The access points and curved ramps to the car park have been dramatised with giant images from Australian films, by graphic designers Emery Vincent Design . The images, in strong primary colours, are screen printed onto perforated metal sheets attached to the external steel stair towers and glass lift shafts. This assists car park users in way-finding and the perennial problem of identifying exactly where their car is parked. Charcoal grey painted concrete upstands, and a palette of primary colours used on columns elevates the design above the purely commercial. The notion of a dingy space inside the car park is completely dispelled with generous two-way access aisles and clear directional signage.

HASSELL has created an elegant solution to a much-maligned typology that has habitually been given little attention. They have managed to balance architectural integrity with the histrionic demands of the entertainment industry.

Opposite and Above: The bold gridded western façade is dramatised with giant images from Australian movies.

Opposite: The northern façade of the car park expresses an elegant solution to a much-maligned typology.

Below: Strong primary colours contrast with the rigorous steel grid of the parking levels.

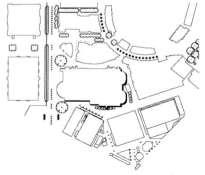

site plan

SYDNEY / NEW SOUTH WALES / AUSTRALIA / 1997–1999

Bent Street Entertainment Precinct

HASSELL's team led by Ken Maher, was 'master architect' for Fox Studios at Moore Park, a project that developed the old Sydney Showgrounds into a major film and entertainment precinct. Initially engaged to provide urban design advice, the brief was extended to design and implement the public domain in the new Studio Tour, Bent Street precinct and Show Ring and to integrate the best of the old showground character with authentic contemporary architecture. The project involved integrating the design work of a number of leading designers and architects. HASSELL also designed the new car park and undertook the design development and documentation of the Studio Tour experience.

The urban form was influenced by the previous site uses and a deliberate diversity is achieved through the careful juxtaposing of new buildings with contrasting architectural character. This strategy, together with the contemporary architectural expression, achieves an environment that successfully avoids any sense of a contrived entertainment precinct.

Below: Market canopies within the showring provide filtered light and geometric patterns.
Opposite: The layout of the pedestrian streets makes reference to earlier site uses.

HONG KONG / PEOPLES REPUBLIC OF CHINA / 1999

Virgin Atlantic Clubhouse

HASSELL was commissioned to design the Virgin Atlantic Clubhouse which provides upper class clubhouse facilites within the Chep Lap Kok Airport.

The new facilities sit on open platforms within the interior volume of the ground departures hall. The design of these club type facilities respects the reductive nature of the Airport architecture, yet introduces warm materials and light effects to give a sense of intimacy to the lounge.

Slatted timber screens and back-lit stone cladding are used with a strict geometric discipline, which results in an underlying connection between architecture and interior design.

Left: Facilities are provided in clearly defined architectural compositions sitting on open platforms.
Opposite: Backlit alabaster panels and finely detailed steel framing lend a richness to the club interior.

OLYMPIC PARK / SYDNEY / NEW SOUTH WALES / AUSTRALIA / 1998–2000

Newington Housing

Parallels can be drawn between the Newington Housing Project at the Sydney Olympic Village (1998-2000) and the Weissenhof Siedlung in Stuttgart (1927). Bruce Eeles, working with HPA Architects for the Mirvac/Lend Lease Development Consortium, developed a Master Plan for the Olympic Village which required that selected architectural firms should work within a strict Modernist language.

HASSELL was appointed to design two linear blocks of apartments in the Village. These explore an interlocking geometry, emphasised with a limited palette of colours. The junction of the apartment blocks and the sky is softened by the introduction of pergolas and louvres. The use of colour in the design distinguishes HASSELL's project from the other apartment blocks. HASSELL was also closely involved in the landscape design associated with the Olympic Village Housing.

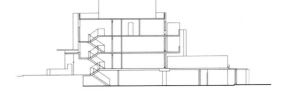
section

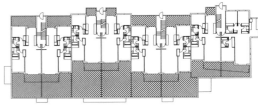
floor plan

Right: Garden terraces and an articulated skyline provide a horizontal scale to the edge of the adjacent wetlands.

Opposite: Highly modelled forms animate the streetscape and give expression to the individual apartments within a cohesive composition.

BRISBANE / QUEENSLAND / AUSTRALIA / 2000

Starcom Worldwide Offices

Starcom Worldwide relocated to the Smellie Building, a heritage listed building which has undergone recent extensive renovation. The existing timber block 'wash down' space in the entry has been preserved and the wonderful arched window on the upper level has become the backdrop to the communal library area. The informal spaces throughout the office such as the 'fun bar' become a central hub for social interaction. The corridor screens contrast with the workstation environment. They provide a cool break out, conceal utilitarian functions and are accented with points of glowing coloured light. The use of light and luminous finishes integral to these spaces provides a somewhat ephemeral quality to the space.

Right: Sinuous furniture elements occupy the atrium floor.

Below: The framing of the mezzanine level provides articulation to the interior space.

Opposite: An inserted stair and lift enliven the central void.

ADELAIDE / SOUTH AUSTRALIA / AUSTRALIA / 1997–2000

David Jones Department Store

HASSELL has a long association with the owners of the David Jones Department Store, having designed their Rundle Mall flagship outlet in the 1970s.

The new David Jones store in Adelaide Central Plaza, completed in 2000, is a high quality retail complex reinforcing David Jones' position as the premier store in the South Australian capital. The store occupies half of a city block, and has two façades, one addressing the pedestrianised Rundle Mall, the other facing the South Australian Museum forecourt (with landscape design by HASSELL) and the new State Library on North Terrace. It is flanked by Charles Street.

The North Terrace façade is elegant and understated, befitting its location on the 'cultural boulevard' of Adelaide. Bronze leaf sculpture by Catherine Truman sweeps across the façade as a reference to the boulevard. The height and detailing of the façade continues a strong 'edge' to the street.

The Rundle Street Mall elevation has a similar grid though more austere, without the addition of sculpture. An intimate and elegant covered street to the east of the store links Rundle Street Mall and North Terrace.

Right: The covered street linking North Terrace and Rundle Mall is enlivened by the ceiling detail and architectural lighting.

Opposite: The North Terrace façade is carefully articulated by expressing the structural grid, and by utilizing a tripartite composition of base, body and roof elements.

Overleaf: The skyline is animated by the separation of cantilevered roof planes. Catherine Truman's artwork decorates the façade.

SYDNEY / NEW SOUTH WALES / AUSTRALIA / 2000

Haymarket Public Spaces

The Haymarket improvements were the result of the Lord Mayor Frank Sartor's vision to develop Sydney as a 24-hour city, with a critical mass of residents and a high quality public realm. A mix of parks, plazas, streets with more cafes and restaurants, and many arts and cultural facilities.[1]

HASSELL worked on the Chinatown area with public artist Peter McGregor and lighting designer Barry Webb. Initially this involved discussions with restauranteurs, market traders and local theatre management. There was naturally some apprehension among the community that the improvements intended to reverse its rundown appearance might inadvertently rob the area of its bustling and chaotic character.

The design concept involved creating a series of external rooms in the north/south streets, contrasting with more directional marking of the east/west streets. The intersection of the streets were marked by major lighting sculptures representing the five essential Chinese elements. The intention was to build on and to enliven the sense of calligraphy and illumination in the precinct.

The project is largely interpretative, and consultant Howard Choy evolved a *feng shui* diagram of Chinatown. Red, green and yellow banners of light stretch across the north-south streets, creating an illuminated ceiling while the smaller cross streets are decorated with colourful box lanterns displaying Chinese calligraphy. The main intersection of Dixon Street Mall is celebrated with an overhead installation, a hovering dome of fibre optics named 'Heaven'. The lighting operates in dynamic programmed cycles to create a theatrical sculptural experience.

 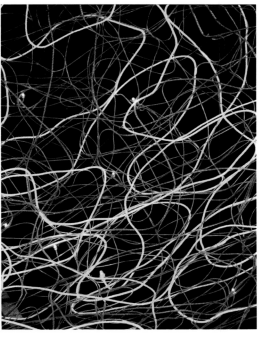

Above Left: Banners of light stretch across Sussex Street.
Above Right and Opposite: 'Heaven', a hovering dome of fibre optics marks the main intersection of the Dixon Street Mall.

[1] Helen Lockhead, 'Sydney Afresh', in *Architecture Australia*, Vol. 88, September/October, 1999. (p72).

ADELAIDE / SOUTH AUSTRALIA / AUSTRALIA / 1999–2000

Roma Mitchell Arts Centre

Opposite: The Roma Mitchell Arts Centre employs red brick, profiled steel and off-form concrete to produce a building that is friendly and eloquent yet tough, hard-edged and resilient.

Named after Australia's first woman judge and former Governor of South Australia, the building is prominently sited in the north-west corner of Light Square. This is a conscious move from Adelaide's 'cultural boulevard' of North Terrace, towards the west end. This bold repositioning of culture is reflected in the building, which is intended to interact with 'ordinary' people as well as the cultural elite.

The project has an unfinished 'rough' quality, concerned with process rather than product. It is building which permits growth and change. Internal finishes include textured stainless-steel wall panels, off-form concrete, galvanised steel handrails, cement screeds and rubber floors. Externally the main material is fair-faced red brick (which makes a connection with the neighbouring heritage-listed Goldsbrough-Mort building), textured stainless-steel panels and steel windows. Shot-blasted concrete paving runs from Light Square into the building at ground floor level.

It is a building for the arts that has no pretensions as a 'High Art' building. It has a 'grunge' quality imparted by the use of 'raw' materials, appropriate for its function as a centre for the contemporary arts, including theatre, dance, fine art and media arts.

The plan and section employ shifting and interlocking geometries. While the rear of the block follows the orthogonal geometry of the city grid, the facade facing Light Square is punctuated by the two theatres, which are rotated and protrude into the square. Clad in stainless steel, they boldly announce the function of the building.

There is a spectacular central atrium linking all floors of the building, which anticipates informal interaction between students and staff. Staircases, balconies and decks are arranged in a seemingly chaotic manner around the central concrete lift shaft and stainless-steel service ducts.

The arrangement of stairs and balconies contributes to the theatrical ambience, and impromptu 'performances' and meetings are stimulated and encouraged. Rails are incorporated throughout the atrium space to accommodate stage lighting for performances, and to display artwork. Students have started to personalise the building. The architects intended that there should be a connection with the street, and that the Centre should interface with the community with activities in the evening spilling out to the street.

The new building makes extensive use of red brick and metal cladding, and one reviewer has noted the allusion to the Modernist and Constructivist origins of Stirling and Gowan's influential Leicester University Engineering Building, UK (1963) which appropriately also derived its composition from surrounding industrial forms.[1]

---

[1] Scott Drake, 'Urban Arts', in *Architecture Australia*, July/August, 2001. (pp 76–79).

Left: The Arts Centre is embedded within the city grid, yet opens on to Light Square. Its distinctive form is an intelligent response to this location.

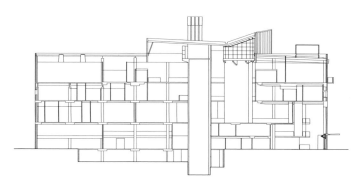

section

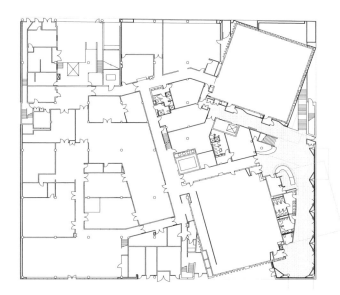

plan

Opposite, Left and Above: This building for the arts has no pretensions, with a 'grunge' quality imparted by the use of 'raw' materials. The arrangement of stairs and balconies contributes to a theatrical ambience, and impromptu 'performances' are encouraged and stimulated.

Overleaf: The plan and section of the building employ shifting and interlocking geometries.

ADELAIDE / SOUTH AUSTRALIA / AUSTRALIA / 2000–2001

Riverbank Promenade

The Torrens River has been the focus for several projects by HASSELL over a period of 30 years. In 1969 John Morphett was commissioned to design the Adelaide Festival Centre located on the river bank, and in 1975 HASSELL was commissioned to master plan the River Torrens Linear Park, a 32 kilometre linear park stretching from the Adelaide Hills to St. Vincents Gulf.

The latest project completed by HASSELL is the 11 metre wide Riverbank Promenade. The catalyst for the master planning of the area now known as the Riverbank Precinct was the State Government's decision to extend the Adelaide Convention Centre. The area between the city's 'cultural boulevard' of North Terrace and the River Torrens contains a concentration of Adelaide's iconic buildings including the State Parliament, Adelaide Railway Station, the Hyatt Regency Hotel and Elder Park in addition to the Festival Centre and the Convention Centre.

Despite the importance of these buildings and institutions, they had been developed over time in an ad-hoc manner and most were introspective. The entire precinct turned its back on the river and the parkland frontage, and divided the city centre from the River Torrens, despite a natural slope which faced north and potentially could be utilised to maximise views over the water body and parkland to the North Adelaide ridge.

In 2000, a master plan was developed for the precinct involving a collaboration between Foster and Partners and HASSELL.

HASSELL, with a design team led by Ken Maher, was selected as the lead consultant for Stage 1 of the elevated east-west pedestrian promenade built over Festival Drive, which links the Festival Centre and the Adelaide Convention Centre.[1] The promenade overlooks the river and terminates 'fingers' of development extending to the river from the city.

Paving and street furniture details include stainless steel and glass balustrades, custom-made signage and stone cladding for planter beds. The aim was to produce an urbane civic setting as a backcloth to people's activities. Large planter boxes contain *ficus hilli* (Hill's Fig) trees.

Careful selection of the street lighting (designed by Renzo Piano/ Meccanoo) makes it a spectacular walkway at night. A new glass elevator (a similar design to that used in the Olympic Park Railway Station) connects the promenade with the car parking and a grand *escalier* links the promenade to the upper plaza levels at the Hyatt Regency and the Convention Centre.

Left: The new promenade successfully reconnects the city with the river, from which it had gradually become estranged.
Overleaf: An urbane civic setting has been created, making it a spectacular walkway at night.

[1] HASSELL was initially involved with the Riverbank Precinct Issue Papers in 1998, and then they led the Riverbank Precinct Master Plan in 1999, and finally led the more detailed Riverbank Precinct External Spaces Study of 2000, which identified a series of projects for staged implementation.

MELBOURNE / VICTORIA / AUSTRALIA / 2001

Jones Lang LaSalle

Real estate services and investment management firm Jones Lang La Salle's brief for the design of their new Melbourne offices was inextricably tied to their changing workplace culture. Their office was seen as a place for networking with clients and colleagues across a diverse portfolio of services and areas of expertise.

Arriving at Jones Lang LaSalle immediately challenges any previously established perception of the organisation as simply an 'Estate Agent'. The company is involved in both the property and recruitment market.

HASSELL's role as interior designer was to project the new global brand for the company while being respectful of established client relationships and the associations with the local brand. There was a need to create an office environment that reinforces the perception of Jones Lang LaSalle as a 'trusted advisor'.

As one steps from the elevator into the dimmed light of the lobby, graphics in the form of coloured diodes and the new company logo projected on the wall react to the visitor's arrival, and indicate lateral movement towards the reception desk.

A visitor is drawn towards a brightly illuminated area and, passing through the reception area, is led to the so-called 'client interface facility' which has been consciously focused outwards, over the skyline of Melbourne. The meeting rooms and lounges are slightly elevated, and a neutral palette of colours is enlivened by original art works.

The office has also been designed to stimulate the corporate vision of the firm as the 'the employer of choice'. The client envisaged their ideal working environment as a collective community, a place where people are encouraged to interact and evolve as a working team.

The flexible workplace has been designed to eliminate boundaries and facilitate open communication. The personnel and size in each team can be adjusted to meet changing scenarios. Each staff member is a node in a number of simultaneous networks that cross business units and roles. There are no dedicated enclosed offices, but there are information and quiet rooms distributed throughout the office.

Central to the success of the design is a large common 'break out' area that is linked between floors by a central staircase. It is a key informal interaction space, a place in which to create social relationships through direct communication.

The former offices of Jones Lang LaSalle were "like a Melbourne Old Boys Club with small offices and a hierarchical management structure". The new offices are an evolutionary step in the client's growth and present a stylish, refined and sophisticated working image. It required HASSELL to have a vision of what "the workplace of the future" might be.[1]

Left: The 'client interface facility' is set back from the building perimeter, creating a circulation gallery to view the skyline form. The meeting spaces are designed as independent pavilions inserted into the building shell.

[1] Tim Shannon, Managing Director of HASSELL in discussion with the author, October 29th, 2002.

Opposite: The research lounge adjacent to reception is utilised as a waiting area, but can be enclosed by sliding glass walls to create a lounge for more informal client meetings.

Below: An aerial view to the central 'breakout' hub from the stair connecting the two levels. The hub forms the social heart of the organisation.

BRISBANE / QUEENSLAND / AUSTRALIA / 1998–2001

Hall Chadwick Centre

The Hall Chadwick Centre was the first medium-rise commercial office tower to be erected in Brisbane in ten years. A proactive role by HASSELL, who developed a model of a climatically responsive building in an urban situation, convinced developer Forrester Kurtz that it was a sound business proposition.

Christopher Wren, the director in charge of the Brisbane office, trained at the Department of Landscape Architecture and Planning at the University of Pennsylvania and brings this background to an office culture that lays emphasis on ecologically sound design.

High-rise buildings in Australia use up to 30% of Australia's grid power. The 22 storey office building was consequently driven by an ecological agenda to use solar power in a cost effective way. It received A$800,000 in research grants and is designed to be a low energy building. The building façade incorporates extensive sunshading and high performance IG glazing units, which result in low solar and UV gain within the offices.

The building makes extensive use of solar panels to generate electricity, and photovoltaic panels are integrated with the roof of the office tower. The panels generate 80-megawatt hours electricity per year, which is 6% of the buildings total electrical demand. Some of the solar power is being used by tenants via an Uninterruptible Power Supply (UPS) and some is being sold back to the grid.

The building design incorporates a whole range of energy efficient technologies including twin air handling units, variable speed fans, high-efficiency chillers, night purge economised cycles and energy saving lighting - which also ensures no glare on computer screens. The energy saving techniques ensures that green house emissions are reduced by 1600 tonnes of CO_2 per year.

The operating costs of the design were worked out in minute detail and energy savings were calculated at A$200,000 per annum. The tower has been acclaimed as the first 4.5 star rated energy saving building in Queensland.

The office tower is located at the corner of Edward Street and Charlotte Street, and has a very large floorplate of 963 square metres. The Brisbane city grid is 100m x 100m but there are planning requirements governing the design of street corners and setbacks, so that all corners finish up being chamfered. There are tough planning constraints on alignment. In this sense planning requirements have driven the design, and not all these requirements were sensible when working in the environmental design mode.

There is a strong theme to the work of the Brisbane office. Design is overlaid by a sense of social and environmental responsibility and the desire, to quote Christopher Wren "to achieve more with less". The fitout of HASSELL's own office within the Hall Chadwick Centre is further evidence of the environmentally responsible design ethos.

Opposite: The elegant corner tower has been acclaimed as the first real energy efficient office building in Queensland.

section

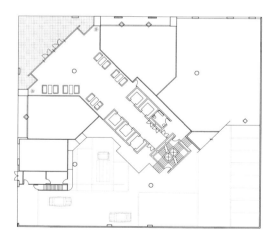

podium plan

Right: The building façades incorporate extensive sunshading and high performance glazing units.

Below: The tower is located at the intersection of Edward Street and Charlotte Street, and the chamfered corners of the building reflect the tough local planning requirements.

Above and Opposite: HASSELL's offices within the tower adopt a minimal aesthetic with maximum transparency and connection to the adjacent roof terraces.

SYDNEY / NEW SOUTH WALES / AUSTRALIA / 2000

Swiss Re Offices

The retrofitting of an office suite on two levels, for Swiss Re Australia, presented an opportunity to create a local interpretation of the company's international corporate identity.

The fitout is intended to reinforce a new workplace culture, following the merger and co-location of two businesses. Fostering teamwork was the prime aim of the design, and this was interpreted as creating additional meeting spaces, especially the encouragement of chance and serendipitous meetings across teams. This also created fewer walls, an emphasis on open space and the abandonment of private offices in order to discourage isolated working.

The design utilises an interconnecting staircase as a central interaction zone. Radiating from this are work zones and support spaces which facilitate team based workshopping, training and innovation sessions, as well as areas for concentrated work. Furniture has been selected to signify these different functions. All spaces are wired for IT giving maximum flexibility in working arrangements.

Opposite: An open stair links the main foyer to the work spaces above.

Right: The café and training area features a mural on the internal wall.

SYDNEY / NEW SOUTH WALES / AUSTRALIA / 1997–2001

North Sydney Olympic Pool

Above: A new café with adjacent terraces is contained by simple planar roof.

Opposite: The new pavilion engages with the urban grain at Milsons Point and with the magnificent view of Sydney Harbour.

The North Sydney Olympic Pool, a 50 metre outdoor pool designed by Rudder and Grout, was opened in April 1936 and used for the 1938 British Empire Games. It is located at Milsons Point at the north end of Harbour Bridge, alongside the iconic Luna Park, which opened to popular acclaim in October 1935. Luna Park is part of Sydney folklore, and it was conceptualised by Herman Phillips and Rupert Browne in the spirit of New York's Coney Island.[1]

But the North Sydney Pool, unlike its exuberant neighbour, is not principally about fun.[2] It has always been a pool for serious swimmers, and in its 65-year history an astounding 86 world records have been set here. The average daily attendance at the pool is 1000, and at any one time dozens of competitive swimmers can be observed ploughing aggressively up and down the lanes, exhorted by stopwatch-holding coaches.

Many of these regular users would have been happy for the antiquated pool to remain unchanged, but time was beginning to take its toll and the facilities were due for a major upgrade. A brief for a two-stage national competition, including a new 25 metre indoor leisure pool, was drawn up and attracted entries from all over Australia. The winners were HASSELL with an innovative yet sensitive design that integrates the new pool with the heritage precinct.

The winning design team, headed by Ken Maher, questioned the design brief and relocated the new indoor pool from a site alongside the main pool to an elevated terrace on the escarpment above the complex.

It is always difficult to judge as to what extent one should 'bend the rules' in a competition, but this was a remarkably astute response to the site conditions, creating a series of platforms stepping down to the bay. The new pool takes the form of an ethereal glazed pavilion, hovering over the existing outdoor pool and reflecting the sky. There is an incredible feeling of lightness and transparency about the new structure. From the interior of the new pool there are stunning views of the waterfront, particularly at dusk.

The project was subject to stringent economic and technical conditions, which resulted in the winning scheme being severely pruned before going ahead, but it has a sophisticated new restaurant at the western end overlooking the original pool. Below the restaurant is a state-of-the-art gym, which also has views of the pool, and there is a sauna, spa, splash pool and a Hall of Fame gallery.

The elegant roof structure of the new indoor pool, with its integrated light slots, is cantilevered out and floats above the existing raked grandstand seating and speaks of integration of old and new.

The street which ran immediately behind the old grandstand is now incorporated into the new building, and the original stair hall with an inserted glass lift shaft provides the circulation link between the new and original pools. Thus architecture provides a palimpsest, embracing layers of memory and meaning.

---

1 Ken Maher and HASSELL collaborated with Paul Berkemeier and McConnell Smith & Johnson in a major restoration of Luna Park in the mid 1990s. The project was awarded the Lachlan Macquarie Award for Conservation by the Royal Australian Institute of Architects in December 1995. The Luna Park restoration was commended as "exemplary in its execution of the conservation process and the quality of work is impressive ...in its use of oral history as well as documentary evidence."
Architecture Australia, Volume 84, Number 6, November/December 1995.

2 Elizabeth Farrelly, 'Ribbons of Light', in *Architecture Australia*, May/June, 2001. (pp 44–47)

Right and Overleaf: The new internal space reflects high technology. Materials are refined and calm, in contrast to the idiosyncratic textures of the existing building.

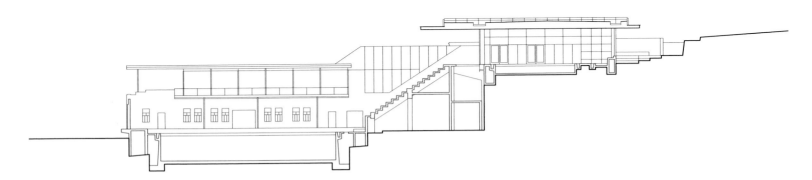

section

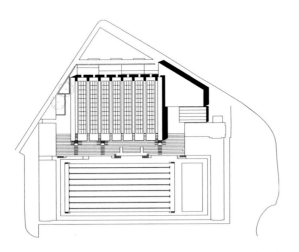

site plan

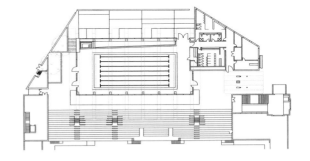

plan

Below: The pool water is heated by a system that extracts heat from the waters of Sydney Harbour and from solar hot water panels mounted on the roof of the new indoor pool.

Opposite: The new 25-metre pool, with its roof covered in solar collectors, is clearly distinguished from the original 50 metre outdoor pool.

GUANGZHOU / PEOPLES REPUBLIC OF CHINA / 2001–2003

Green Garden Clubhouse

Generous clubhouse facilities and high quality design combine to establish a unique contemporary living environment, reflecting the changing aspirations of the new China housing market. This is one of a number of HASSELL projects in which early planning and design stages integrate the design of interior facilities, and establish a unique image and character within each development. Green Garden in Guangzhou has pursued a strong sense of sophisticated modernity where the interior spaces and facilities engage with the landscape setting.

Above, Right and Opposite: A spare approach to spaces and materials gives a sophisticated quality to the reception, bar and library. Consistency of detailing has been adopted to fully integrate the exterior and interior environment.

HONG KONG / PEOPLES REPUBLIC OF CHINA / 2001–2002

London Mission Clubhouse

Located at the western Mid Level of Hong Kong Island, the 19th Century London Mission Building (1893) was originally the dormitory of the Nethersole Hospital and is listed grade 3 by the Antiquity and Monuments Ordinance (AMO) as a preserved historical building in Hong Kong. The scope of the project converted the British Victorian-style London Mission Building to an exclusive Residents' Clubhouse, accommodating Function Rooms, Library, Snooker Room and Health Club where residents gather and organize social events.

The Residents' Clubhouse establishes an extension of the living room environment, with sophisticated and flexible dining and functional facilities in an annex to the residential development at 80 Robinson Road. The Clubhouse establishes a distinctive character reinforcing the building's Victorian heritage and contrasting with a simple contemporary interior.

Above: Restored façade of the British Victorian-style building.

Right and Opposite: Contemporary interiors are seamlessly inserted into the original fabric.

BAROSSA VALLEY / SOUTH AUSTRALIA / AUSTRALIA / 2001

Jacobs Creek Visitor Centre

Johann Gramp, an immigrant of German origin, founded Orlando Wines in 1847, and planted one of the first vineyards in the Barossa Valley on the banks of Jacobs Creek. The function of the new Visitor Centre is to market the products of the now internationally renowned vineyard. Although the client was familiar with similar centres, which utilise traditional architectural language, the brief requested a modern building using modern technology.

Nevertheless, HASSELL questioned the original site as it offered limited scope, and scoured the area to identify an alternative, which, due to its location in the Jacobs Creek Reserve, had to go through the non-complying process. The planning procedure took over a year. The selected site is spectacular and looks over the rolling wine country of the Barossa Valley. The entrance road passes through a ford at Jacobs Creek, which creates a poetic link with the early pioneers. The restaurant, which was added late in the design process, has proved to be a major visitor attraction.

The Visitor Centre is an elegant building, a glazed box with a wafer-thin steel roof supported on minimal steel columns projecting far beyond the box. Floor to ceiling windows with narrow horizontal transoms give maximum transparency, and the building's low profile ensures that it integrates with the landscape. HASSELL's design is environmentally sensitive, and sustainability was in the forefront of their thinking. They wanted to attract native birds and to encourage native plants, and dozens of gum trees were planted to re-establish this native species. The project uses wastewater treatment and rainwater harvesting. Rainwater is used for flushing toilets while wastewater is treated and used for watering the gum trees and for cleaning purposes.

Timber used for the structure of the Visitor Centre was recycled from a bridge in Queensland. There are reconstituted timber veneers on the bar front, and the flooring is built from bamboo strips.

The building is naturally ventilated, employing steel panels which are heated by the sun and which encourage convection. Louvres automatically open at high level to encourage cross ventilation, and extensive glass walls are shaded by wide overhanging roofs. Overall, the Visitor Centre is an excellent example of HASSELL's ecologically responsible approach to design.

Left: Floor to ceiling windows provide maximum transparency and integration with the landscape.

section

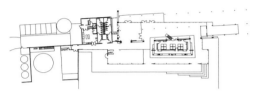
plan

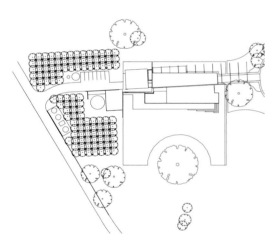
site plan

Right: The special relationship of the Visitor Centre to the landscape of the Barossa Valley was the outcome of HASSELL's exhaustive search for 'the right site'.

Left and Below: The Jacobs Creek Visitor Centre is an object in the landscape that is modest and unassuming yet distinctive and stylish.

BRISBANE / QUEENSLAND / AUSTRALIA / 2002

Gadens Offices

The interior design of Levels 24 and 25 of the Commonwealth Bank Building in Brisbane was consequent upon the merger of Gadens with another law firm, Bowdens.

Offices are of equal size and all workstations are standardised to permit maximum flexibility in staff arrangements. The offices are cellular, but glass walls are employed and glass sliding doors create a feeling of openness. The staff room and a training room are linked by an internal stair, and are located on an exterior wall with excellent views. A spacious entrance lobby has been created on Level 25 with panoramic views over the Brisbane River.

To symbolise the revitalised and progressive new partnership, HASSELL designed a light filled space with a neutral palette of colours as a backcloth to the firm's extensive art collection. Warm white is the dominant colour in both the public and private areas.

Opposite: Strong colour, such as the coloured glass on the internal stair, is applied in straregic areas.

Above Right: The staff breakout areas are linked by an internal stairs. With an emphasis on an environment that encourages social interaction, these spaces also support informal meetings.

Below Right and Overleaf: The simplicity of the fitout as a backdrop allows the artworks to make a strong statement.

UNIVERSITY OF ADELAIDE / SOUTH AUSTRALIA / AUSTRALIA / 2000–2002

Santos Petroleum Engineering Building

The Santos Petroleum Engineering Building is located within a master plan prepared by MGT Architects for the University of Adelaide campus, and adjoins a block of similar height, materials and proportions designed by MGT.

The Engineering Building designed by HASSELL is of sturdy construction, its brick clad columns and façade being faintly reminiscent of Louis Kahn's seminal Indian Institute of Management in Ahmedabad (1963). A single storey arcade runs along the façade of the block alongside the central turfed courtyard and links with the MGT building.

HASSELL's project is slightly angled away from the earlier block by MGT and this entasis gives added drama to the main axis of the central courtyard, which terminates at the Barr Smith Library building.

Right and Opposite: A vertical emphasis within a sturdy cubic composition is provided by the glazed stairway.

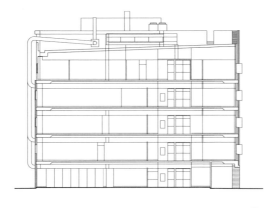
section

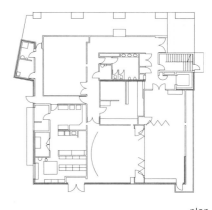
plan

Right: An urbane yet accessible character is provided by the main street frontage.

Opposite: Masonry elements define the perimeter to a courtyard facing the historic Barr Smith Library. Strip windows are carefully detailed to enhance the solid modelling of the façade.

BRISBANE / QUEENSLAND / AUSTRALIA / 1997–1998

Tognini's Salon

 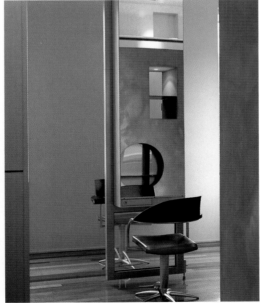

Tognini's is known in the hair industry as an innovative, cutting edge salon. The new studio is a direct interpretation of the culture of Tognini's, expressing the distinctiveness of the proprietors. Metallic surfaces of bronze, stainless steel, wire, mesh and concrete create a clean, modern and efficient atmosphere. A unique feature is a group of networking tables wired for personal television and laptops, or for talking, reading and relaxing. Proposed by Tognini's as a workshop for the busy urban professional, the brief for the project was more complex than a straightforward studio. Within a difficult narrow space, varied procedures, from 'facials' to 'fullday total body treatments' had to be accommodated. The design has produced a clean, cool and well-lit space with a restrained palette of colour and material.

Opposite: Cement rendered product walls serve many functions and screen a wash basin zone from the more visible styling area.

Above left: A wall of bronze and silver wire mesh guides clients to the reception area.

Above right: Customised 4 metre high cutting stations set into a grid track system, allow linear, angled or staggered placement.

BRISBANE / QUEENSLAND / AUSTRALIA / 2000–2002

Margate Foreshore Redevelopment

Won in a 1994 Australian Institute of Landscape Architects (AILA) national urban design competition, the HASSELL scheme for the Margate Foreshore Development has transformed a 1.5 kilometre stretch of sea front between Suttons Beach and Woody Point. Formerly dominated by vehicles, the waterfront has been turned into an asset enjoyed by a wide cross section of the community.

The project includes a three metre wide boardwalk, tree planting, pavilions to give shelter and commanding views, access for cyclists and pedestrians, dune restoration, rerouting of overground services, car parks and traffic calming measures. Pollutant traps have been installed to improve water quality and appearance. Future stages will include the establishment of strategic headlands to stabilise erosion and create facilities for recreation.

Nine white-painted timber pavilions have been introduced along the beachfront. They are simple contemporary structures designed to fit the character of the area. They give shade and shelter, and help create an attractive pedestrian environment. In character they are not unlike the Victorian bathing huts that once lined the beach.

The pavilions are located at the intersection of side streets with the foreshore, so that they function as meeting points for local residents as well as visitors. A range of timber seats and picnic tables has been designed to complement the pavilions.

Right: The transformation of the Margate beachfront has created something special from the ordinary and everyday.

SYDNEY / NEW SOUTH WALES / AUSTRALIA / 2000

Maroubra Beach Foreshore Promenade

Maroubra Beach is one of southern Sydney's best kept secrets, and recognised as one of the city's premier surfing beaches. In 1998 Randwick City Council approached HASSELL to provide detailed design and documentation for a major improvement scheme to revitalise the Maroubra Beach area.

The project included upgrading the existing foreshore promenade and seawall, a new amenities block built into the sandstone cliff face, new ramps and stairs facilitating universal beach access, a new skate park facility for the area's youth, and an upgrade of the Marine Parade streetscape and pavement.

A number of specially commissioned artworks prepared by three separate artists further enliven the foreshore promenade, including tile and pebble mosaics and specialist feature lighting.

Right: Ramps and broad steps provide a transition between the beach and the promenade.

BEIJING / PEOPLES REPUBLIC OF CHINA / 2001–2003

Goldfield Green Town

HASSELL provided the master planning for this 24ha residential development in Beijing, and designed the landscape for all exterior spaces.

A series of blade walls and steel columns with formal planting, and a similarly formal use of water elements give a strong architectural character to the whole development.

Apartment blocks are located on the edge of the site, and provide enclosure for the central garden space. The layout allows for visual links throughout, which take advantage of the open space to the west and northeast while allowing landscape and visual corridors to the north and south-east. Careful consideration has been given to the surrounding sites, allowing integration of block height with the adjacent developments.

The landscape concept establishes an ordered framework for the open spaces, using a combination of stone, brickwork and rendered walls to provide variety within a cohesive precinct.

Opposite and right: The stone walls, steel framework and formal planting soften the impact of the large development, and introduce a strong architectural character.

SYDNEY / NEW SOUTH WALES / AUSTRALIA / 2000

Hassell Sydney Office

HASSELL's new Sydney office fitout creates a workplace with a range of spatial experiences. The unusually open office design places the entire organisation clearly on view and celebrates the culture of the HASSELL team. The use of a simple palette of black, white and grey reinforces the range of formality found in the meeting and social spaces, all of which seem to be hard-working activity-based areas.

ESD principles have been employed to reduce energy consumption and indoor air quality while providing an enhanced, people focused workplace in keeping with HASSELL's long-standing commitment to environmental considerations.

A central 'street' acts as an organising element through the workplace and is a common focus for staff and visitors.

The interior design of the space maximises the volumetric variations achievable by changing ceiling heights, contrasting light and dark colours and a mix of furniture that plays up the differing levels of formality in the common spaces. The element of surprise is ever-present with new vistas opening up as the viewer moves around the floor, yet there is a cohesive openness to the entire space.

Top Right: Ceiling panels are removed from the central street to emphasise its form.

Bottom Right: A zinc clad wall acts as a backdrop to an informal, linear reception space.

Allens Arthur Robinson Offices

This workplace for a leading law practice provides a strong identity and a standard of excellence based on contemporary work patterns. Simple and carefully composed architectural elements with high quality finishes are used to give a sense of spaciousness and order.

The reception area provides the most formal conference room, defined by the bronzed aluminium battened acoustic ceiling. Four meeting rooms to the west are hotel-like in materials, lighting and detailing. Adjacent is a large multipurpose zone that can be divided into four separate conference rooms, or a number of variations from lecture mode to high-level audio visual and video conferencing facilities.

The remaining areas are characterised by less formal spaces such as lounges, where the ceiling heights are lowered to create a domestic scale. These breakout spaces are enclosed by large sliding glass walls that, when left open, connect the circulation path to external landmarks and views beyond.

Opposite: The view from a corner node back to the reception area.
Top Right: The waiting area and lift lobby have stained oak panelling on the walls and ceilings.
Below Right: The formal conference room has views through to the reception area.
Overleaf: A dramatic spiral stair slices through the reception and waiting area.

SYDNEY / NEW SOUTH WALES / AUSTRALIA / 1999–2003

Victoria and Tote Parks

Created by Sir James Joynton Smith in 1908, the Victoria Park Race Track was a popular attraction for Sydney racegoers for almost half a century, but by 1958 horse racing had been supplanted by industry. Viscount Nuffield and the British Motor Company (BMC) opened what was acknowledged to be the most modern car manufacturing plant in the Southern Hemisphere. The Department of Naval Defence Stores used other parts of the site.

This use would last for less than 40 years and in 1997, ownership of the site in Zetland passed to developers Landcom, who engaged HASSELL to prepare a refined master plan of the 25-hectare site with the intention of creating a sustainable urban community. Following the preparation of the master plan, a limited design competition was held for the design of the parks and public domain. The proposals for the public domain involved a strong environmental ethos, with propositions for water recycling systems. HASSELL was invited to undertake the project in collaboration with the NSW Department of Public Works and Services (DPWS), with HASSELL focusing on the parks and the DPWS undertaking the streets.

By 2003 the major infrastructure including Woolwash, Tote and Joynton Parks was completed, with over 300 apartments and terrace houses occupied. Joynton Park, the largest of the four public open spaces, is situated in the centre of the development and has a spectacular water cascade, designed by artists Jennifer Turpin and Michaelie Crawford.

HASSELL in collaboration with architects Stanisic Turner, have also prepared the larger Green Square Master Plan covering the suburbs of Zetland, Beaconsfield and parts of Waterloo, Alexandria and Rosebery - an area of 275 hectares. The plan establishes a mix of housing types for 25,000 people and accommodation for service and high-tech industries.

Opposite: A shimmering water cascade, designed by artists Jennifer Turpin and Michaelie Crawford, edges Joynton Park.
Top Right: Informally placed timber benches recall the previous industrial site character.
Bottom Right: An assymetrical geometry underlies Tote Park.

SYDNEY / NEW SOUTH WALES / AUSTRALIA / 1999–2002

The National Institute of Dramatic Art

Opposite: The soaring foyer and 'veil' of aluminium louvres are theatrically illuminated at night.

The National Institute of Dramatic Art (NIDA) has launched the careers of some of Australia's most talented actors, directors, set designers, costume designers and writers. Mel Gibson, Baz Luhrmann, Judy Davis, Kate Blanchard and Pamela Stephenson are alumni of NIDA.

The original 1985 NIDA complex was designed by Peter Armstrong of the Government Architects Office. It is an introverted, collegiate type of building with a 'wall' of single storey brick administrative offices fronting Anzac Parade. Another row of two storey rectangular brick structures containing rehearsal rooms and a small black-box theatre are separated from the offices by a series of courtyards. These courtyards, which incorporate steps and curved seating, serve as outdoor performance spaces. Essentially, Stage 1 of NIDA was an academic institution without a substantial presence in the public domain.

It was always planned to build a second stage incorporating a large major performance venue, to provide the public face of NIDA, and with support from the Centenary of the Federation Fund this got under way in 1999.

The NIDA management were familiar with the design of the Canberra Playhouse Theatre and admired it as a performance venue. For this reason they approached HASSELL when they decided to proceed with the Parade Theatre.

The theories of the British theatre designer Iain McIntosh had been influential in the design of the Canberra building. McIntosh advocates a return to the Shakespearean theatre form with balconies and a reduced distance to the stage, as the theatres of the 20th Century had moved away in spatial terms from traditional theatre. Sight lines had become the principal design criteria and, in the process, auditoria had lost their intimacy. In the traditional form some of the sight lines are inevitably compromised, but there is more engagement of the audience with the actors.

HASSELL worked with Peter Armstrong to ensure that the old and the new complement each other. The new building is extroverted to complement the introverted character of the original, the new is light and transparent compared with the old, which is solid, and the new is accessible to the public compared with the private nature of the original.

However any suggestion that these are two completely different buildings is quickly dispelled when one enters the foyer which is sited precisely at the point at which the two buildings join.[1] The idea of old and new and the juxtaposition of the two is simply perfect.

But the real success of the new theatre is in the intimacy of the auditorium. Having sat in the rear row of the upper circle for a performance of a one-woman show by Bea Arthur, I can testify to the excellent acoustics and the feeling of being part of the performance. 750 seats (Canberra has 600) are arranged in semi-circular fashion around the auditorium drum and on its raked floor, with no member of the audience more than 17 metres from the stage.

The client was enthusiastic about the notion of the 'richness' of theatre; they wanted warmth and a sense of conviviality. The colours of the seats range from warm pink and orange at the centre to red and

[1] Michael Ostwald, Performing Art, in *Monument*, No. 50, August/September, Terraplanet Limited, Sydney, 2002. (pp 48-54)

mauve at the periphery. The gradation of colours ensures that when there is a small audience concentrated in the centre, the darker periphery seats fade into the shadow. The upper galleries are fronted by preformed Tasmanian blackwood panels which step up and around the curve following the line of the seating.

NIDA is a semicircular auditorium enclosed in a glazed rectangular box, with the intervening space used as a foyer and for vertical circulation. The metaphor of an 'egg in a box' could be used to describe this form. A semicircular skylight bridges between the perimeter of the drum and the horizontal plane of the roof.

An elaborate 'veil' of aluminium louvres supported on steel ribs is suspended from the semicircular ring beam that supports the edge of the roof. The veil simultaneously obscures and reveals, visually enhancing the auditorium drum. Illuminated from above by sunlight during the day, and dramatically enhanced by coloured spotlights at night, the foyer space and its staircases become a public performance venue.

The veil protrudes through the roof and creates a distinctive silhouette visible to pedestrians and commuters on Anzac Parade. The louvres, which are almost closed at the top of the veil, are nearly horizontal at the base. They serve a practical purpose, channelling chilled air into the foyer from the plenum and daylight into the foyer.

At an urban level, the challenge was how to relate to Anzac Parade and the strong axis of the University of New South Wales Master Plan directly across the road. The client was keen to have a strong presence on Anzac Parade and to signal independence from the University. When viewed through the 11-metre high glazed curtain wall that fronts the theatre, the foyer becomes a highly dramatised extension of the public domain. Lights in the projecting canopy preclude the need for street lighting poles, and the street is drawn into the inner space.

Left and Below: The rich materiality and intimacy of the auditorium (left) complements the dramatic and flamboyant space of the foyer (below).

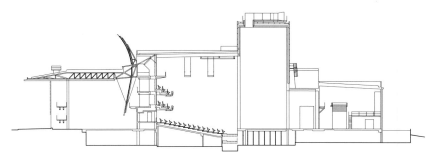

section

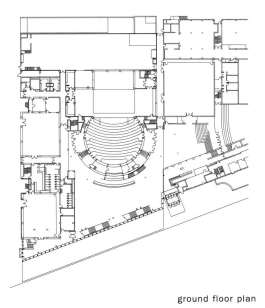

ground floor plan

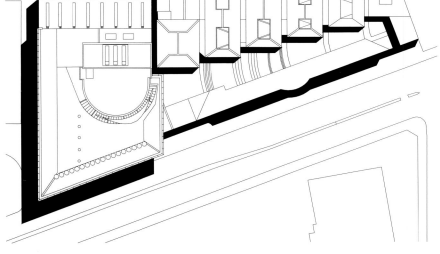

site plan

Above and Opposite: The foyer spaces and the staircase can provide a theatrical space for public performance.

Left and Opposite: The new foyer is extroverted to complement the introverted character of the original 1985 building.

Overleaf: The foyer, when viewed through the high glazed curtain wall, becomes a dramatised extension to the public domain.

Awards

Since its inception, one of the principal objectives of the Firm has been to maintain the highest standard of design. This has been rewarded by success in planning & design competitions and by the presentation of design awards for building projects of widely varying characters. The following is a list of recognition for contribution of excellence within the planning and design industry.

Gadens Lawyers Office Fitout, Brisbane, QLD
2003 Royal Australian Institute of Architects (Qld Chapter) Awards - High Commendation, Category - Interior Architecture

2003 Royal Australian Institute of Architects (Qld Chapter) Regional Awards – Brisbane Regional Commendation, Category – Interior Architecture

2003 Design Institute of Australia (Qld Chapter) Awards – Award of Merit, Category – Interior Design (Corporate)

Hall Chadwick Centre, 120 Edward Street, Brisbane, QLD
2003 The Architecture Show Magazine and The Francis Greenway Society Green Building Awards – Silver Medal

2003 Property Council of Australia National Awards – National Rider Hunt Award

2003 Property Council of Australia (Qld Chapter) Awards – Rider Hunt Award for Queensland

2003 Australian Property Institute (Qld Chapter) Excellence in Property Awards – Property Development Award

Victoria Park Public Domain, Zetland, NSW
2003 Royal Australian Institute of Architects (NSW Chapter) Awards – Lloyd Rees Award, Category – Civic Design

2003 Royal Australian Institute of Architects (NSW Chapter) Awards – World Environment Day ESD Award

2003 Cement and Concrete Association of Australia National Public Domain Awards – Commendation, Category – Precinct

2002 Green Square Design Awards – Award Winner, Category – Innovation, for the design of an environmentally sustainable public domain

2002 The Architecture Show Magazine and The Francis Greenway Society Green Building Awards – Gold Medal

2002 Australian Institute of Landscape Architects (NSW and ACT State Groups) Awards – Project Award, Category – Master Planning

2001 Urban Development Institute of Australia (NSW Division) Awards – Commendation, Category – Professional Consultancy (in association with NSW Department of Public Works and Services

University of Adelaide, Santos Petroleum Engineering Building, Adelaide, SA
2003 Royal Australian Institute of Architects (SA Chapter) Awards – Commendation Award, Category – New Building

Jacob's Creek Visitor Centre, Barossa Valley, SA
2003 Royal Australian Institute of Architects (SA Chapter) Awards - Award of Merit, Category - Commercial

2003 Royal Australian Institute of Architects (SA Chapter) Awards - Commendation Award, Category - Interior Architecture

2003 Royal Australian Institute of Architects (SA Chapter) Awards – Commendation Award, Category – Sustainable Architecture

2002 Design Institute of Australia (SA Chapter) Awards - Commendation

111 James Street Apartment, New Farm, QLD
2003 Design Institute of Australia (Qld Chapter) Awards – Award of Merit, Category – Interior Design (Residential)

2003 Design Institute of Australia (Qld Chapter) Awards – Commendation, Category – Furniture Design

Simpson and Owen Residence, Ascot, QLD
2003 Design Institute of Australia (Qld Chapter) Awards – Commendation, Category – Interior Design (Residential)

Tognini's Café, Spring Hill, QLD
2003 Dulux Colour Awards - Qld State Commendation

'Dunavant Cotton' Headquarters, QLD
2002 Design Institute of Australia (Qld Chapter) Awards – Commendation, Category – Interior Design (Corporate)

Elizabeth Street Estate, Richmond, VIC
2002 Planning Institute Australia (Victorian Chapter) Awards – Certificate of Commendation, Category – Community Planning

Regional Workforce Accommodation Solutions Study, SA
2003 Planning Institute of Australia National Awards – Merit, Category – Rural & Regional Planning Achievement

2002 Planning Institute of Australia (SA Division) Awards - Award for Excellence, Category – Rural & Regional Planning Achievement

Towards a Strategic Urban Plan for the City of Dili, East Timor
2003 Planning Institute of Australia National Awards – President's Award

2002 Planning Institute of Australia (SA Division) Awards - Award for Excellence, Category - Community Planning

Eyre Peninsula Tourism Development Strategy 2002-2006, SA
2002 Planning Institute of Australia (SA Division) Awards – Certificate of Commendation, Category - Rural & Regional Planning Achievement

Buckland Park & Environs Study, SA
2002 Planning Institute of Australia (SA Division) Awards – Certificate of Commendation, Category - Urban Planning Practice (Concept)

Cooks Cove Master Plan, Sydney, NSW
2002 Australian Institute of Landscape Architects National Awards - Merit Award, Category - Master Planning

HASSELL Office, Brisbane, QLD
2002 Royal Australian Institute of Architects (Qld Chapter) Awards – Commendation Award, Category – Interior Architecture

National Institute of Dramatic Art, Sydney, NSW
2002 The Architecture Show Magazine and The Francis Greenway Society Green Building Awards – Gold Medal

2002 Royal Australian Institute of Architects (NSW Chapter) Awards - Sir John Sulman Award for Architectural Merit

The Age Print Centre Project, Tullamarine, VIC
2002 Royal Australian Institute of Architects (Victorian Chapter) Awards – Sir Osborne McCutcheon Award, Category - Commercial

Margate Foreshore Redevelopment Project, QLD
2002 Australian Institute of Landscape Architects National Awards - Merit Award, Category – Design - Public Spaces

2002 Australian Institute of Landscape Architects (Qld Group) Awards – Harry Oakman Tribute Project Award, Category – Local Government Design

2002 Australian Institute of Landscape Architects (Qld Group) Awards – Merit Award, Category – Open Space and Recreational Design

ACI Public Domain Plan, NSW
2002 Green Square Design Awards – Award Winner, Category – Greening, for the design of public spaces and linear parks

Metropolitan Greenspace Program Advisory Services, Sydney, NSW
2002 Australian Institute of Landscape Architects (NSW Group) Awards – Merit Award, Category – Research and Communication

2001 NSW Premier's Public Sector Awards – Bronze Award, Category – Best Practice and Business Management

2000 Australian Institute of Landscape Architects National Project Awards - Project Award of Merit, Category – Research and Communication

Kelvin Grove Urban Village, Brisbane, QLD
2002 QLD Premier's Award for Excellence in Public Sector Management – Award Winner, Category – Sustainable Environment

2002 QLD Premier's Award for Excellence in Public Sector Management – Commendation, Category – Innovation and Creativity

2002 Australian Institute of Landscape Architects National Awards - Project Award, Category – Master Planning

2002 Royal Australian Institute of Architects (Qld Chapter) Awards – High Commendation, Category - Unbuilt

2002 Royal Australian Planning Institute National Awards – Urban Planning Achievement Award

2001 Royal Australian Planning Institute (Qld Division) Awards – Award of Excellence, Category – Urban Planning Achievement

2001 Royal Australian Planning Institute (Qld Division) Awards – Certificate of Merit, Category – Urban Design

The Lilydale Project, VIC
2002 Planning Institute Australia (Victorian Chapter) Awards – Certificate of Commendation, Category – Urban Design – Plans and Ideas

2001 Australian Institute of Landscape Architects (Victorian & Tasmanian State Groups) Awards - Merit Award, Category – Landscape Planning - Urban Master Plans

Blinman Town Centre Plan, SA
2001 Royal Australian Planning Institute (SA Division) Awards – Award of Excellence, Category – Rural and Regional Planning Achievement

Dry Areas Consultation and Research, SA
2001 Royal Australian Planning Institute (SA Division) Awards – Certificate of Commendation, Category – Community Planning

Port Lincoln Action Plan, SA
2001 Royal Australian Planning Institute (SA Division) Awards – Certificate of Commendation, Category – Urban Planning Practice (Concept)

South East Catchment Water Management Plan, SA
2001 Royal Australian Planning Institute (SA Division) Awards – Certificate of Commendation, Category – Environmental Planning or Conservation

Canberra Playhouse, ACT
2001 Property Council of Australia (ACT Division) Awards - Rider Hunt Award

Roma Mitchell Arts Education Centre, Adelaide, SA
2001 Royal Australian Institute of Architects (SA Chapter) Awards – Award of Merit, Category – New Building

2001 The Adelaide Prize – Adelaide City Council - Commendation for Excellence

Adelaide Central Plaza (David Jones), Adelaide, SA
2002 Property Council of Australia (SA Chapter) Awards – Rider Hunt Award

2002 Property Council of Australia (National) Awards – Certificate of Merit

2001 Royal Australian Institute of Architects (SA Chapter) Awards – Commendation, Category - Commercial

2001 The Adelaide Prize – Adelaide City Council – Commendation for Excellence

North Sydney Olympic Pool, NSW
2001 Australian Institute of Steel Construction (NSW) Awards –Architectural Steel Design Awards - High Commendation, Category - Architectural Structures

2001 Royal Australian Institute of Architects (NSW Chapter) Awards – Award for Architecture, Category – Public Buildings

2001 The Architecture Show Magazine and The Francis Greenway Society Green Building Awards – Silver Medal

2001 EnergyAustralia National Trust Heritage Awards – Winner Energy Australia Award, Category J

Bendigo City Plan, VIC
2001 Australian Institute of Landscape Architects (Victorian and Tasmanian State Groups) Awards - Commendation Award, Category – Landscape Planning - Policy Development

2000 Royal Australian Planning Institute (Victorian Division) Awards for Planning Excellence – Award Commendation, Category – Urban Design – Plans and Ideas

Fox Studios Car Park, Sydney, NSW
2001 Royal Australian Institute of Architects (NSW Chapter) Awards – Commendation, Category - Commercial Buildings

2000 Dulux Colour Awards - Award Winner, Category – Commercial Exterior

Millennium Parklands Concept Plan/Hill Road Corridor, Sydney, NSW
2001 The Architecture Show Magazine and The Francis Greenway Society Green Building Awards – Gold Medal

2000 Building Design Professions Urban Design Awards – Urban Design in Australia Award

2000 Australian Institute of Landscape Architects National Project Awards - Project Award in Landscape Architecture, Category - Master Planning

2000 Australian Institute of Landscape Architects National Project Awards - Project Award in Landscape Architecture, Category – Design Rehabilitation and Conservation

1999 Australian Institute of Landscape Architects (NSW and ACT State Groups) Awards - Planning Merit Award, Category - Master Planning

Riverbank Precinct Master Plan, Adelaide, SA
2000 Australian Institute of Landscape Architects (SA Group) Awards – State Project Award, Category – Master Planning

Charles Street Upgrade, Adelaide, SA
2000 Australian Institute of Landscape Architects (SA Group) Awards – State Merit Award, Category – Civic Design

2001 The Adelaide Prize – Adelaide City Council – Commendation for Excellence

Fox Studios Parade Ring, Sydney, NSW
2000 Lightweight Structures Association of Australia Design Awards – Citation, Category – Special Applications

Athletics Stadium, Mile End, SA
2000 Metal Building Awards – Finalist, Category – Community – Sporting

Haymarket Priority Design Project, Dixon Street, Chinatown, Sydney, NSW
2000 Australian Institute of Landscape Architects National Project Awards - Project Award of Merit, Category – Design - Landscape Art

Adelaide Park Lands Management Strategy, SA
2000 Australian Institute of Landscape Architects (SA Group) Awards – State Merit Award, Category – Environmental Planning

2000 Australian Institute of Landscape Architects National Project Awards - Project Award of Merit, Category – Master Planning

Next Generation Leisure Centre, Adelaide, SA
2000 Design Institute of Australia Awards
– Commendation Award

Sydney Tropical Centre Redevelopment Feasibility Study, Royal Botanic Gardens, Sydney, NSW
2000 Australian Institute of Landscape Architects National Project Awards - Project Award of Merit, Category – Master Planning

The Frankston Project, VIC
2000 Royal Australian Planning Institute (Victorian Division) Awards for Planning Excellence – Award Commendation, Category – Local Government Planning Process

1999 Australian Institute of Landscape Architects (Victorian and Tasmanian State Groups) Awards - Commendation Award, Category - Master Planning

Cairnlea Residential Estate, VIC
2000 Royal Australian Planning Institute (Victorian Division) Awards for Planning Excellence – Award Commendation, Category – Urban Planning Achievement

2000 Urban Development Institute of Australia (Victorian Division) Awards - Award for Excellence, Category – Environmental

Tennis SA Stadium Upgrade, SA
2000 Design Institute of Australia Awards – Commendation Award

2000 Royal Australian Institute of Architects (SA Chapter) Awards – Commendation Award, Category - Recycling

1999 Australian Institute of Steel Construction (SA) Awards - High Commendation

National Motor Museum, Birdwood, SA
2000 Metal Building Awards – Certificate of Merit, Category – Community - Environment

2000 Property Council of Australia (SA Chapter) Awards – Certificate of Merit

1999 Royal Australian Institute of Architects (SA Chapter) Awards -Commendation Award, Category - New Building

1999 Australian Institute of Steel Construction (SA) Awards - Winner Architectural Steel Design Award

Commonwealth Law Courts, Melbourne, VIC
2000 Australian Institute of Landscape Architects National Project Awards - Project Award of Merit, Category – Design - Public Spaces

2000 Royal Australian Institute of Architects (Victorian Chapter) Awards – Award of Merit, Category – Urban Design

1999 Australian Institute of Landscape Architects (Victorian and Tasmanian State Groups) Awards - Commendation Award, Category - Building Settings

1999 Royal Australian Institute of Architects (Victorian Chapter) Awards - Marion Mahony Award, Category - Interior Architecture

1999 Royal Australian Institute of Architects (Victorian Chapter) Awards - Commendation Award, Category - Institutional New

1999 Dulux Colour Awards - Award Winner, Category - Commercial Interior

QANTAS Domestic Terminal, Sydney, NSW
2000 Royal Australian Institute of Architects (NSW Chapter) Awards – Commercial Building Architecture Award, Category – Commercial Building

1998 Metal Building Awards - Award of Merit, Category - Commercial

1997 H.H. Robertson Awards - Award of Excellence for Outstanding Architectural Design

1997 Australian Institute of Steel Construction (NSW) Awards - Architectural Steel Design Award

Harrington Park Estate, Sydney, NSW
2000 Urban Development Institute of Australia National Awards – Winner Best Community Building

1999 Urban Development Institute of Australia (NSW Division) Awards – Winner Community Estate of the Year

1996 Urban Development Institute of Australia National Awards - Award of Excellence in Urban Development, Category - Residential Development of 200 Lots or Greater

1996 Urban Development Institute of Australia (NSW Division) Awards - Award of Excellence in Urban Development, Category - Residential Development of 200 Lots or Greater

1995 Royal Australian Planning Institute (NSW Division) Awards -Commendation for Excellence in Planning

Ohel Leah Synagogue Conservation Works, Hong Kong
2000 UNESCO Asia-Pacific Heritage Awards for Culture Heritage Conservation - Outstanding Project Award

1999 Hong Kong Institute of Architects Awards - Award

1999 Asia Pacific Interior Design Awards - Blanche Gallardo Award

Tognini's Hair Workshop, Brisbane, QLD
1999 Royal Australian Institute of Architects (Qld Chapter) Awards – Regional Commendation

1999 Hair Expo Award for Best Salon

Territory Wildlife Park Master Plan, NT
1999 Australian Institute of Landscape Architects (Qld Group) Awards - Merit Award, Category – Landscape Master Planning

'M' on the Bund Restaurant, Shanghai, China
1999 Asia Pacific Design Awards, Category - Best Restaurant over 200m2

Gowrie Creek Catchment Management Strategy, Toowoomba, QLD
1999 Royal Australian Planning Institute (Qld Division) Awards – Certificate of Merit, Category – Environmental Planning

Herriott's Glen Estate, VIC
1999 Urban Development Institute of Australia (Victorian Division) Awards - Award for Excellence, Category - Residential Development of 200 Lots or More

Olympic Park Rail Station, Sydney, NSW
1999 Australian Institute of Landscape Architects (NSW and ACT State Groups) Awards – Design Merit, Category - Urban & Civic Design

1999 Australian Institute of Steel Construction (NSW) Awards - Architectural Steel Design Award

1998 Royal Australian Institute of Architects National Awards - Sir Zelman Cowen Award for Public Buildings

1998 Royal Australian Institute of Architects National Awards - Access Citation

1998 Royal Australian Institute of Architects (NSW Chapter) Awards - Sir John Sulman Award for Outstanding Architecture

1998 Royal Australian Institute of Architects (NSW Chapter) Awards - BHP Colorbond Award for the Innovative Use of Steel in Architecture

1998 Metal Building Awards - Award of Excellence

1998 Metal Building Awards - Award, Category - Civil Engineering

Empire Theatre, Toowoomba, QLD
1999 National Trust (Qld) Awards - John Herbert Award for Excellence in Heritage Conservation Works or Actions

1998 Property Council of Australia Awards - High Commendation Award

1997 Royal Australian Institute of Architects (Qld Chapter) Awards - Regional Commendation

1997 Illuminating Engineering Society of Australia and New Zealand Awards - Commendation Award

Chifley Square, Sydney, NSW
1998 Australian Institute of Landscape Architects National Project Awards - Project Award in Landscape Architecture, Category - Design - Public Spaces

SA Water - Australia Asia Water Centre, Adelaide, SA
1998 Royal Australian Institute of Architects (SA Chapter) Awards - Commendation Award

1998 Design Institute of Australia (SA Chapter) Award - Commendation Award

Redlands Indigiscapes Centre Master Plan, QLD
1998 Australian Institute of Landscape Architects (Qld Group) Awards - Merit Award, Category – Landscape Master Plans

Gordon Yu-Hoi Chiu Building, University of Sydney, NSW
1998 Royal Australian Institute of Architects (NSW Chapter) Awards - Award for Architecture, Category - Public Buildings

Bankers Trust Landscape, Laffer's Triangle, SA
1998 Civic Trust of South Australia Inc Awards - Commendation, Category - Landscape and Streetscape

Bankers Trust Australia, Science Park, Adelaide, SA
1998 Royal Australian Institute of Architects (SA Chapter) Awards - Award of Merit, Category - Commercial

City of Adelaide Street Furniture, SA
1998 Civic Trust of South Australia Inc Awards - Commendation, Category - Landscape and Streetscape

Riverview Urban Renewal, QLD
1998 Australian Institute of Landscape Architects National Project Awards - Project Award of Merit, Category – Design - Residential

1997 Australian Institute of Landscape Architects (Qld Group) Awards - Project Award, Category - Residential Design – Broad Scale

Victoria's Open Range Zoo at Werribee, VIC
1998 BHP Metal Building Awards - Certificate of Merit

1997 Australian Institute of Landscape Architects (Victorian & Tasmanian State Groups) Awards - Award of Merit, Category - Open Space and Recreational Design

Torrens Comprehensive Catchment Water Management Plan, SA
1998 Australian Institute of Landscape Architects National Project Awards – Project Award of Merit, Category - Environmental Planning

1997 Stormwater Industry Association Inc Awards - Award for Environmental Excellence

International Terminal, Melbourne Airport, Tullamarine, VIC
1997 Property Council of Australia (Victorian Chapter) Awards - Overall Winner

1997 Property Council of Australia (Victorian Chapter) Awards - Award, Category - Public Buildings

Dandenong Transport Interchange, VIC
1997 Australian Institute of Steel Construction (Victoria) Awards - Architectural Steel Design Award

Garden East, Adelaide, SA
1997 Urban Development Institute of Australia (SA Division) Awards - Award of Excellence in Urban Development, Category – Urban Redevelopment

Tiger Island, Dreamworld, QLD
1997 Australian Institute of Landscape Architects (Qld Group) Awards - Project Award, Category - Open Space and Recreational Design

Rizal Park, The Philippines
1997 Australian Institute of Landscape Architects (Qld Group) Awards - Merit Award, Category - International Projects

Tai Po New Town Waterfront Park, Hong Kong
1997 Hong Kong Institute of Landscape Architects Awards - Award of Merit

Business Excellence in Greater China & Asia
1997 Australian Chamber of Commerce/ANZ Award of Commendation

Desert Wildlife Park and Botanic Gardens Concept and Master Plan, Alice Springs, NT
1997 Australian Institute of Landscape Architects (Qld Group) Awards -Project Award, Category - Judges Special Acknowledgment

1996 Australian Institute of Landscape Architects National Project Awards - Project Award of Merit, Category - Planning

1995 Royal Australian Planning Institute National Awards – Certificate of Merit, Category - Occasional Special Award

Swinburne University of Technology, David Williamson Theatre, Prahran Campus, VIC
1996 Royal Australian Institute of Architects (Victorian Chapter) Awards - Award of Merit, Category - Institutional Alterations and Extensions

River Park Estate, SA
1996 Urban Development Institute of Australia (SA Division) Awards - Award of Excellence in Urban Development, Category - Residential Development of 200 Lots or Greater

St Vincent's Public Hospital, Melbourne, VIC
1996 Royal Australian Institute of Architects (Victorian Chapter) Awards - Commendation Award, Category - Institutional New

Christian Brothers College Eastern Courtyard Redevelopment, SA
1996 Civic Trust of South Australia Inc Awards - Award of Merit, Category - Landscape and Streetscape

North West Adelaide Strategic Plan, SA
1996 Royal Australian Planning Institute (SA Division) Awards - Award for Excellence in Planning, Category - Planning Scholarship

West End Urban Development Strategy, Adelaide, SA
1996 Royal Australian Planning Institute (SA Division) Awards - Award for Excellence in Planning, Category - Community Planning

Jawbone Flora and Fauna Reserve, Williamstown, VIC
1996 Australian Institute of Landscape Architects National Project Awards - Project Award of Merit, Category – Design

Streeton Views Residential Estate, VIC
1996 Urban Development Institute of Australia National Awards - Award of Excellence in Urban Development, Category - Residential Development of 200 Lots or Greater

1995 Urban Development Institute of Australia (Victorian Division) Awards - Award for Excellence in Urban Design, Category - Residential Development of 200 Lots or More

1994 Royal Australian Planning Institute (Victorian Division) Awards - Honourable Mention, Category - Planning and Development

Regent Gardens Housing Project, Northfield, SA
1996 Urban Development Institute of Australia National Awards - Award of Excellence in Urban Development, Category - Residential Development of 200 Lots or Greater

1995 Urban Development Institute of Australia (SA Division) Awards - Award of Excellence in Urban Design, Category
- Residential Development of 200 Lots or Greater

1995 Stormwater Industry Association Inc Awards - Award for Environmental Excellence in Stormwater Management

1995 Royal Australian Planning Institute National Awards - Certificate of Merit, Category - Planning and Development

1994 Royal Australian Planning Institute (SA Division) Awards - Award for Excellence in Planning, Category - Planning and Development

Stormwater Flood Control and Water Quality Improvement, SA
1995 Civic Trust of South Australia Inc Awards - Certificate of Commendation, Category - Landscape and Streetscape

Luna Park, Sydney, NSW
1995 Royal Australian Institute of Architects National Awards - Lachlan Macquarie Award for Conservation

1995 Royal Australian Institute of Architects (NSW Chapter) Awards - Award of Merit, Category - Conservation

1995 Wattle (NSW Country Division) Awards - Award for Colour in Architecture

Para Institute of TAFE, Salisbury Campus, SA
1995 Building Owners and Managers Association (SA Division) Awards – Award

Kensington Banks, VIC
1995 Royal Australian Planning Institute (Victorian Division) Awards for Planning Excellence - Award, Category - Planning and Development

Port of Newcastle, Kooragang Island Industrial Land Study, NSW
1995 Royal Australian Planning Institute (NSW Division) Awards - Commendation for Excellence in Planning

Unley Social Policy and Strategic Plan, SA

1995 Royal Australian Planning Institute (SA Division) Awards -Certificate of Merit, Category - Community Planning

Western Munno Para Master Plan, SA
1995 Royal Australian Planning Institute (SA Division) Awards - Award for Excellence in Planning, Category - Urban Planning Achievement

Elizabeth College of TAFE, Adelaide, SA
1995 Building Owners and Managers Association (SA Division) Awards – Award

St Andrew's Hospital Central Wing Development, Adelaide, SA
1995 Building Owners and Managers Association (SA Division) Awards - Commendation Award

University of South Australia, Amy Wheaton Building, Magill Campus, Adelaide, SA
1995 Royal Australian Institute of Architects (SA Chapter) Awards -Award of Merit

Blackbutt Wildlife Exhibits Stages 1 and 2, NSW
1995 Hunter Region Design Awards - Merit Award for the Design of a Development Incorporating Outstanding Landscape and Urban Design, Category – Open

1995 Landscape Contractors Association of New South Wales Awards - Award for Excellence, Category – Recreation

1994 Hunter Region Design Awards - Alfred Sharp Award for Excellence in the Design of a Development Incorporating Outstanding Landscape and Urban Design, Category - Open

1994 Landscape Contractors Association of New South Wales Awards - Award for Excellence, Category – Open

North Terrace Urban Design Study, Adelaide, SA
1995 Royal Australian Institute of Architects (SA Chapter) Awards - Award for Excellence

1994 Royal Australian Planning Institute (SA Division) Awards - Award for Excellence in Planning, Category - Urban Planning Achievement

Upper Spencer Gulf Housing Study, SA
1994 Royal Australian Planning Institute (SA Division) Awards – Award for Excellence in Planning, Category - Rural Planning Achievement

ETSA, Rymill Park, Adelaide, SA
1994 Royal Australian Institute of Architects (SA Chapter) Awards - Award of Merit, Category - Recycling

1994 Design Institute of Australia Awards - Commendation Award for Interior Design

City of Port Adelaide Recreation and Sport Plan, SA
1994 Royal Australian Planning Institute National Awards - Certificate of Merit, Category - Community Planning

1993 Royal Australian Planning Institute (SA Division) Awards - Award for Excellence in Planning, Category - Community Planning

Barker Inlet Wetlands, SA
1993/94 Institution of Engineers Australia (SA Division) Awards - Engineering Excellence Award, Category – Environment

120 Collins Street Tower, Melbourne, VIC
1994 Master Builders Association of Victoria Awards - Excellence in Construction Award, Category - Office Accommodation and/or Public Buildings Over $5M

1993 Building Owners and Managers Association Awards - National Award

1993 Building Owners and Managers Association (Victorian Division) Awards - Award

1992 Royal Australian Institute of Architects (Victorian Chapter) Awards - Award of Merit, Category - Commercial New

Little Manly Point, Sydney, NSW
1993 Institution of Engineers Australia (NSW Division) Awards - Engineering Excellence Award, Category – Environment

120 Collins Street Tower, Melbourne, VIC
1994 Master Builders Association of Victoria Awards - Excellence in Construction Award, Category - Office Accommodation and/or Public Buildings Over $5M

1993 Building Owners and Managers Association Awards - National Award

1993 Building Owners and Managers Association (Victorian Division) Awards - Award

1992 Royal Australian Institute of Architects (Victorian Chapter) Awards - Award of Merit, Category - Commercial New

Marrickville Recreation User Needs Study, NSW
1993 Royal Australian Planning Institute (NSW Division) Awards - Award for Excellence in Planning

Better Drainage: Guidelines for Multiple Use of Drainage Systems, NSW
1993 Royal Australian Planning Institute (NSW Division) Awards - Award for Excellence in Planning, Category - Planning by State and Federal Government

Travelling Scholarship
1992 Australian Institute of Landscape Architects National Project Awards - Project Award of Merit, Category - Special Initiatives

Darwin Airport, NT
1992 Royal Australian Institute of Architects (NT Chapter) Awards - Corporate and Office Design Award

Coopers and Lybrand Tower, Auckland, New Zealand
1992 New Zealand Institute of Architects National Awards - Award for Architecture

Athelstone and Highbury River Torrens Linear Park and Residential Development, SA
1992 Royal Australian Planning Institute (SA Division) Awards - Award for Excellence in Planning, Category - Metropolitan Planning

River Torrens Linear Park, Adelaide, SA
1992 Australian Institute of Landscape Architects National Project Awards - Project Award in Landscape Architecture, Category – Infrastructure

Bicentennial Conservatory, Adelaide, SA
1992 Australian Institute of Landscape Architects National Project Awards - Project Award in Landscape Architecture, Category - Building Settings

World of Primates, Adelaide Zoo, SA
1992 Australian Institute of Landscape Architects National Project Awards - Project Award of Merit, Category - Parks and Recreation

Broken Hill Screen & Crushing Plant Objection Development Application, NSW
1992 Royal Australian Planning Institute (NSW Division) Awards -Award for Excellence in Planning

Evaluation of the Methodology to Determine Optimum Residential Land Stocks, SA
1992 Royal Australian Planning Institute (SA Division) Awards - Commendation Award for Excellence in Planning, Category - Planning Document

Barossa Valley Region - Development Standard and Guidelines Table, SA
1992 Royal Australian Planning Institute (SA Division) Awards -Commendation Award for Excellence in Planning, Category - Non-Metropolitan Planning

City of Brighton Residential Supplementary Development Plan, SA
1992 Royal Australian Planning Institute (SA Division) Awards - Award for Excellence in Planning, Category - Statutory Document

Finlaysons Interiors, Adelaide, SA
1992 Design Institute of Australia (SA Chapter) Awards - Award of Merit, Category - Commercial Interiors

1991 Royal Australian Institute of Architects (SA Chapter) Awards - Award of Merit, Category - Interiors

Finlaysons Office Building, Adelaide, SA
1991 Royal Australian Institute of Architects (SA Chapter) Awards -Award of Merit, Category – Commercial

Cancer Care Centre, Illawarra Regional Hospital, NSW
1991 Wollongong City Council Design Awards – Certificate of Merit, Category - Public Buildings

City of West Torrens Residential Development Supplementary Development Plan, SA
1991 Royal Australian Planning Institute (SA Division) Awards - Special Commendation Award for Excellence in Planning, Category - Statutory Document

Whyalla Human Services Study, SA
1991 Royal Australian Planning Institute (SA Division) Awards - Award for Excellence in Planning, Category - Community Project

Sir Donald Bradman Stand, SA
1991 Clay Brick Association Awards - Clay Brick Award

1990 Royal Australian Institute of Architects (SA Chapter) Awards -Commendation Award

1990 Civic Trust of South Australia Awards - Award of Merit

Tuen Mun Town Park, Hong Kong
1990 Hong Kong Institute of Landscape Architects Awards - Silver Medal

Adelaide-Crafers Highway Draft Environmental Impact Statement, SA
1990 Royal Australian Planning Institute National Awards - Certificate of Merit, Category - Excellence in Planning Document

1989 Royal Australian Planning Institute (SA Division) Awards - Award for Excellence in Planning, Category - Planning Document

1988 Association of Consulting Engineers of Australia Awards - Award of Merit

Westminster Performing Arts Centre, SA
1989 Royal Australian Institute of Architects (SA Chapter) Awards -Award of Merit

Psychiatric Hostel, Bruce, ACT
1988 Royal Australian Institute of Architects (ACT Chapter) Awards – The Canberra Medallion for Architectural Merit

Enterprise House Office Building, SA
1988 Royal Australian Institute of Architects (SA Chapter) Awards -Commendation Award, Category - Commercial New

Aldermans Solicitors, Adelaide, SA
1988 Royal Australian Institute of Architects (SA Chapter) Awards -Commendation Award, Category - Interior Design

Wollongong City Council Administration and Library Centre, NSW
1987 Wollongong City Council Building Design Awards - Certificate of Merit, Category - Public Building

Adelaide 2000 Urban Design Competition, SA
1986 Royal Australian Institute of Architects (SA Chapter) Awards -Design Mention Award

River Torrens Linear Park and Flood Mitigation Scheme, Adelaide, SA
1986 Civic Trust of South Australia Inc Awards – Civic Trust Award

River Torrens Study, Adelaide, SA
1986 Australian Institute of Landscape Architects National Project Awards - Award in Landscape Architecture, Category – Research and Studies

The Craigburn Estate, SA
1986 Civic Trust of South Australia Inc Awards - Award of Merit, Category - Town Planning

Noarlunga Town Plaza, SA
1986 Australian Institute of Landscape Architects National Project Awards - Project Award of Merit, Category - Civic Design

1986 Clay Brick Association Awards - Clay Brick Award

1977 Civic Trust Awards - Certificate of Merit

1977 Clay Brick Association Awards - Clay Brick Award

Technology Park, Adelaide, SA
1985 Greening of Adelaide Recognition Award

Sir Samuel Way Building, Adelaide, SA
1984 Royal Australian Institute of Architects National Awards - Lachlan Macquarie Award Finalist

1984 Royal Australian Institute of Architects (SA Chapter) Awards -Commendation Award

Wangal Centenary Bushland Reserve, NSW
1984 Landscape Contractors Association of New South Wales Awards - Award for Most Environmentally Beneficial Landscape Area

The Colonel Light Centre, Adelaide, SA
1982 Civic Trust of South Australia Inc Awards - Certificate of Commendation

Leigh Creek South Town Centre, SA
1982 Civic Trust Awards - Certificate of Merit

Rundle Car Park, Adelaide, SA
1980 Royal Australian Institute of Architects (SA Chapter) Awards -Award of Merit

Adelaide Festival Theatre, SA
1974 Royal Australian Institute of Architects (SA Chapter) Awards - Award of Merit

1974 Timber Development Association – Award

Balm Paints Administration, Rocklea, QLD
1966 Royal Australian Institute of Architects Awards - Building of the Year

Studios for Channel 'O' Austarama Television Pty Ltd, VIC
1965 Royal Australian Institute of Architects (Victorian Chapter) Awards - Architecture Medal

House, SA
1963 Timber Development Association Awards - House of the Year Award

Factory Building for H J Heinz, Dandenong, VIC
1955 Architecture and Arts Award

Acknowledgements

The publishers would like to thank Laraine Sperling, Linda Fraser and Elizabeth Kumm of HASSELL for their assistance in the production of this book.

The line drawings were produced by Alice Martin, Tam Nguyen and Elisa Sutanudjaja.

Photography Credits

All photography © by Patrick Bingham-Hall except as © listed below

Page 14 (Both Pictures) – Architectural Photographics Ltd

Page 15 (Top) – Arnold Studios Ltd

Page 15 (Bottom) – Don Meller

Page 16 (Both Pictures) – Tom Balfour

Page 17 – Chester Ong

Page 20-21 – Chester Ong

Page 23 (Bottom) – Goldfield (Beijing) Honeyeah Development Co Ltd

Page 26 – Philip Hyam

Page 27 (Centre and Bottom) – HASSELL/C3D

Page 28 (Both Pictures) – HASSELL/C3D

Page 29 – HASSELL/C3D

Page 38 (All Pictures) – HASSELL (Christopher Wren)

Page 39 – Neil Lorimer

Page 41 – John Gollings

Page 43 – John Gollings

Pages 46 – Chester Ong

Pages 50-51 – Trevor Mein

Pages 75-77 – Chester Ong

Page 93 (Left) – Martin Saunders

Page 93 (Right) – John Gollings

Page 94 – John Gollings

Pages 96-97 – Trevor Mein

Pages 106-107 – Chester Ong

Pages 110-111 – David Sandison

Page 117 – HASSELL

Page 126 – Steve Rendoulis

Page 128 – Steve Rendoulis

Pages 130-133 – Trevor Mein

Pages 138-139 – Vincent Long

Pages 140-141 – Tyrone Brannigan

Pages 150-151 – Chester Ong

Pages 153 – Chester Ong

Page 159 – Steve Rendoulis

Pages 160-163 – David Sandison

Page 166 – Steve Rendoulis

Pages 168-169 – David Sandison

Page 175 – Scott Sothers

Pages 176-179 – Earl Carter

Page 181 – HASSELL